Introduction to Circuits and Systems

Introduction to Circuits and Systems

Edited by
Jasper Harrison

Larsen & Keller
www.larsen-keller.com

Introduction to Circuits and Systems
Edited by Jasper Harrison
ISBN: 978-1-63549-067-1 (Hardback)

© 2017 Larsen & Keller

Larsen & Keller

Published by Larsen and Keller Education,
5 Penn Plaza,
19th Floor,
New York, NY 10001, USA

Cataloging-in-Publication Data

Introduction to circuits and systems / edited by Jasper Harrison.
     p. cm.
Includes bibliographical references and index.
ISBN 978-1-63549-067-1
1. Electronic circuits. 2. Electric circuits. 3. Electrical engineering.
4. Electronics. I. Harrison, Jasper.
TK7867 .I58 2017
621.381 5--dc23

The publisher's policy is to use permanent paper from mills that operate a sustainable forestry policy. Furthermore, the publisher ensures that the text paper and cover boards used have met acceptable environmental accreditation standards.

Printed and bound in the United States of America.

For more information regarding Larsen and Keller Education and its products, please visit the publisher's website www.larsen-keller.com

Table of Contents

Preface

This book is a compilation of chapters that discuss the most vital concepts in the field of circuits and systems. It discusses the various technologies that have been developed in this field of study. Circuits and systems provide networks and channels for movement of electrical energy, signals and data and information as well as provide connection between disparate electrical components. Selected concepts that redefine this area of circuits and systems have been presented in this text. It explores all the important aspects of the field in the present day scenario. This textbook aims to serve as a resource guide for students and facilitate the study of the discipline.

A short introduction to every chapter is written below to provide an overview of the content of the book:

Chapter 1 - Electronic circuits consist of resistors, transistors and capacitors. Electronic circuits can easily be categorized as analog circuit, digital circuit or mixed-signal circuit. This chapter will provide an integrated understanding of electronic circuit; **Chapter 2** - Integrated circuits are sets of electronic circuits that are present in one small chip. The following section focuses on integrated circuit development, semiconductor device fabrication, wafer-scale integration, three-dimensional integrated circuit, op amp integrator etc. This chapter is an overview of the subject matter incorporating all the major aspects of integrated circuit; **Chapter 3** - Linear circuits are electronic circuits; a better definition of linear circuits that can be given is that it follows the superposition principle. The section strategically encompasses and incorporates the major components and key concepts of linear circuit, providing a complete understanding; **Chapter 4** - The various circuit theorems discussed in this text are ports, generators, Norton's theorem, Miller's theorem and extra element theorem. The topics discussed in the chapter are of great importance to broaden the existing knowledge on circuit theorems; **Chapter 5** - Electronics is a field of science that has electricity as its most fundamental part. Some of the features of electronics elucidated in this text are digital electronics, analogue electronics, electronic circuit design and electronic components. The aspects elucidated in this section are of vital importance, and provides a better understanding of electronics; **Chapter 6** - The field of engineering that studies the functions of electricity, electronics and electromagnetism is known as electrical engineering. Electronic engineering particularly concentrates on devices, microprocessors and microcontrollers. The following text helps the readers to understand all the allied fields of electronic circuit.

Finally, I would like to thank my fellow scholars who gave constructive feedback and my family members who supported me at every step.

Editor

Introduction to Electronic Circuit

Electronic circuits consist of resistors, transistors and capacitors. Electronic circuits can easily be categorized as analog circuit, digital circuit or mixed-signal circuit. This chapter will provide an integrated understanding of electronic circuit.

An electronic circuit is composed of individual electronic components, such as resistors, transistors, capacitors, inductors and diodes, connected by conductive wires or traces through which electric current can flow. The combination of components and wires allows various simple and complex operations to be performed: signals can be amplified, computations can be performed, and data can be moved from one place to another.

The die from an Intel 8742, an 8-bit microcontroller that includes a CPU, 128 bytes of RAM, 2048 bytes of EPROM, and I/O "data" on current chip.

Circuits can be constructed of discrete components connected by individual pieces of wire, but today it is much more common to create interconnections by photolithographic techniques on a laminated substrate (a printed circuit board or PCB) and solder the components to these interconnections to create a finished circuit. In an integrated circuit or IC, the components and interconnections are formed on the same substrate, typically a semiconductor such as silicon or (less commonly) gallium arsenide.

An electronic circuit can usually be categorized as an analog circuit, a digital circuit, or a mixed-signal circuit (a combination of analog circuits and digital circuits).

Breadboards, perfboards, and stripboards are common for testing new designs. They allow the designer to make quick changes to the circuit during development.

A circuit built on a printed circuit board (PCB).

Analog Circuits

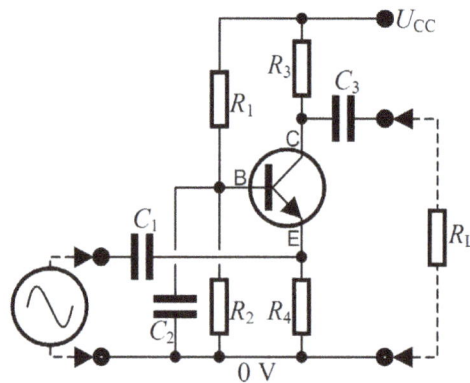

A circuit diagram representing an analog circuit, in this case a simple amplifier

Analog electronic circuits are those in which current or voltage may vary continuously with time to correspond to the information being represented. Analog circuitry is constructed from two fundamental building blocks: series and parallel circuits. In a series circuit, the same current passes through a series of components. A string of Christmas lights is a good example of a series circuit: if one goes out, they all do. In a parallel circuit, all the components are connected to the same voltage, and the current divides between the various components according to their resistance.

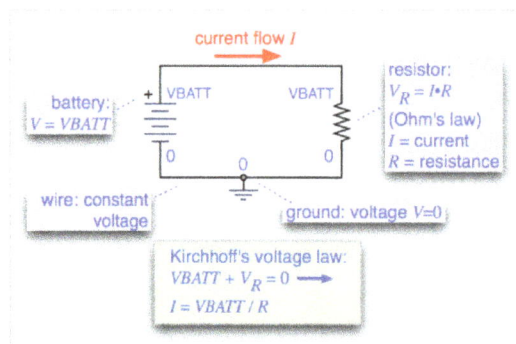

A simple schematic showing wires, a resistor, and a battery

The basic components of analog circuits are wires, resistors, capacitors, inductors, diodes, and transistors. (In 2012 it was demonstrated that memristors can be added to the list of available components.) Analog circuits are very commonly represented in schematic diagrams, in which wires are shown as lines, and each component has a unique symbol. Analog circuit analysis employs Kirchhoff's circuit laws: all the currents at a node (a place where wires meet), and the voltage around a closed loop of wires is 0. Wires are usually treated as ideal zero-voltage interconnections; any resistance or reactance is captured by explicitly adding a parasitic element, such as a discrete resistor or inductor. Active components such as transistors are often treated as controlled current or voltage sources: for example, a field-effect transistor can be modeled as a current source from the source to the drain, with the current controlled by the gate-source voltage.

When the circuit size is comparable to a wavelength of the relevant signal frequency, a more sophisticated approach must be used. Wires are treated as transmission lines, with (hopefully) constant characteristic impedance, and the impedances at the start and end determine transmitted and reflected waves on the line. Such considerations typically become important for circuit boards at frequencies above a GHz; integrated circuits are smaller and can be treated as lumped elements for frequencies less than 10 10GHz or so.

An alternative model is to take independent power sources and induction as basic electronic units; this allows modeling frequency dependent negative resistors, gyrators, negative impedance converters, and dependent sources as secondary electronic components

Digital Circuits

In digital electronic circuits, electric signals take on discrete values, to represent logical and numeric values. These values represent the information that is being processed. In the vast majority of cases, binary encoding is used: one voltage (typically the more positive value) represents a binary '1' and another voltage (usually a value near the ground potential, 0 V) represents a binary '0'. Digital circuits make extensive use of transistors, interconnected to create logic gates that provide the functions of Boolean logic: AND, NAND, OR, NOR, XOR and all possible combinations thereof. Transistors interconnected so as to provide positive feedback are used as latches and flip flops, circuits that have two or more metastable states, and remain in one of these states until changed by an external input. Digital circuits therefore can provide both logic and memory, enabling them to perform arbitrary computational functions. (Memory based on flip-flops is known as static random-access memory (SRAM). Memory based on the storage of charge in a capacitor, dynamic random-access memory (DRAM) is also widely used.)

The design process for digital circuits is fundamentally different from the process for analog circuits. Each logic gate regenerates the binary signal, so the designer need not account for distortion, gain control, offset voltages, and other concerns faced in an analog design. As a consequence, extremely complex digital circuits, with billions of logic elements integrated on a single silicon chip, can be fabricated at low cost. Such digital integrated circuits are ubiquitous in modern electronic devices, such as calculators, mobile phone handsets, and computers. As digital circuits become more complex, issues of time delay, logic races, power dissipation, non-ideal switching, on-chip and inter-chip loading, and leakage currents, become limitations to the density, speed and performance.

Digital circuitry is used to create general purpose computing chips, such as microprocessors, and custom-designed logic circuits, known as application-specific integrated circuit (ASICs). Field-programmable gate arrays (FPGAs), chips with logic circuitry whose configuration can be modified after fabrication, are also widely used in prototyping and development.

Mixed-signal Circuits

Mixed-signal or hybrid circuits contain elements of both analog and digital circuits. Examples include comparators, timers, phase-locked loops, analog-to-digital converters, and digital-to-analog converters. Most modern radio and communications circuitry uses mixed signal circuits. For example, in a receiver, analog circuitry is used to amplify and frequency-convert signals so that they reach a suitable state to be converted into digital values, after which further signal processing can be performed in the digital domain.

Integrated Circuit: An Overview

Integrated circuits are sets of electronic circuits that are present in one small chip. The following section focuses on integrated circuit development, semiconductor device fabrication, wafer-scale integration, three-dimensional integrated circuit, op amp integrator etc. This chapter is an overview of the subject matter incorporating all the major aspects of integrated circuit.

Integrated Circuit

An integrated circuit or monolithic integrated circuit (also referred to as an IC, a chip, or a microchip) is a set of electronic circuits on one small flat piece (or "chip") of semiconductor material, normally silicon. This can be made much smaller than a discrete circuit made from independent electronic components - a modern chip may have several billion transistors in an area the size of a human fingernail. Over the past half century, the size, speed, and capacity of chips has increased enormously, driven by technical advances that allow more and more transistors on chips of the same size. These advances, collectively known as Moore's law, allow a computer chip of 2016 to have a million times the capacity and a thousand times the speed of the initial computer chips of the early 1970s. Integrated circuits are used in virtually all electronic equipment today and have revolutionized the world of electronics. Computers, mobile phones, and other digital home appliances are now inextricable parts of the structure of modern societies, made possible by the low cost of ICs.

Erasable programmable read-only memory integrated circuits. These packages have a transparent window that shows the die inside. The window allows the memory to be erased by exposing the chip to ultraviolet light.

ICs were made possible by experimental discoveries showing that semiconductor devices could perform the functions of vacuum tubes and by mid-20th-century technology advancements in semiconductor device fabrication. The integration of large numbers of tiny transistors into a small chip was an enormous improvement over the manual assembly of circuits using discrete electronic components. The integrated circuit's mass production capability, reliability and building-block approach to circuit design ensured the rapid adoption of standardized integrated circuits in place of designs using discrete transistors.

ICs have two main advantages over discrete circuits: cost and performance. Cost is low because the chips, with all their components, are printed as a unit by photolithography rather than being constructed one transistor at a time. Furthermore, packaged ICs use much less material than discrete circuits. Performance is high because the IC's components switch quickly and consume little power (compared to their discrete counterparts) because of their small size and close proximity. Over the years, transistor sizes have decreased from 10s of microns in the early 1970s to around 14 nanometers in 2014 with a corresponding million-fold increase in transistors per unit area. As of 2016, typical chip areas range from a few square millimeters to around 600 mm², with up to 25 million transistors per mm².

Terminology

An *integrated circuit* is defined as:

A circuit in which all or some of the circuit elements are inseparably associated and electrically interconnected so that it is considered to be indivisible for the purposes of construction and commerce.

Circuits meeting this definition can be constructed using many different technologies, including thin-film transistor, thick film technology, or hybrid integrated circuit. However, in general usage *integrated circuit* has come to refer to the single-piece circuit construction originally known as a *monolithic integrated circuit*.

Invention

Early developments of the integrated circuit go back to 1949, when German engineer Werner Jacobi (Siemens AG) filed a patent for an integrated-circuit-like semiconductor amplifying device showing five transistors on a common substrate in a 3-stage amplifier arrangement. Jacobi disclosed small and cheap hearing aids as typical industrial applications of his patent. An immediate commercial use of his patent has not been reported.

The idea of the integrated circuit was conceived by Geoffrey W.A. Dummer (1909–2002), a radar scientist working for the Royal Radar Establishment of the British Ministry of Defence. Dummer presented the idea to the public at the Symposium on Progress in Quality Electronic Components in Washington, D.C. on 7 May 1952. He gave many symposia publicly to propagate his ideas, and unsuccessfully attempted to build such a circuit in 1956.

A precursor idea to the IC was to create small ceramic squares (wafers), each containing a single miniaturized component. Components could then be integrated and wired into a bidimensional or tridimensional compact grid. This idea, which seemed very promising in 1957, was proposed to the US Army by Jack Kilby and led to the short-lived Micromodule Program (similar to 1951's

Project Tinkertoy). However, as the project was gaining momentum, Kilby came up with a new, revolutionary design: the IC.

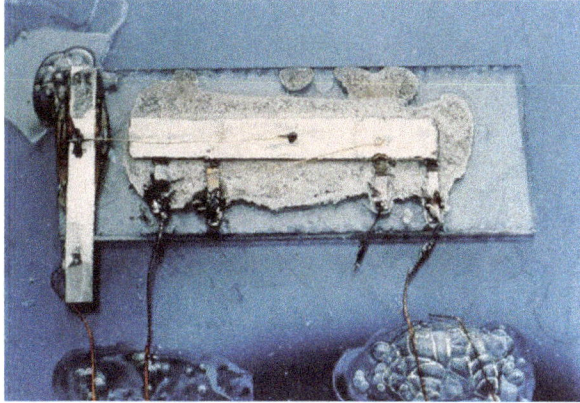

Jack Kilby's original integrated circuit

Newly employed by Texas Instruments, Kilby recorded his initial ideas concerning the integrated circuit in July 1958, successfully demonstrating the first working integrated example on 12 September 1958. In his patent application of 6 February 1959, Kilby described his new device as "a body of semiconductor material ... wherein all the components of the electronic circuit are completely integrated." The first customer for the new invention was the US Air Force.

Kilby won the 2000 Nobel Prize in Physics for his part in the invention of the integrated circuit. His work was named an IEEE Milestone in 2009.

Half a year after Kilby, Robert Noyce at Fairchild Semiconductor developed his own idea of an integrated circuit that solved many practical problems Kilby's had not. Noyce's design was made of silicon, whereas Kilby's chip was made of germanium. Noyce credited Kurt Lehovec of Sprague Electric for the *principle of p–n junction isolation* caused by the action of a biased p–n junction (the diode) as a key concept behind the IC.

Fairchild Semiconductor was also home of the first silicon-gate IC technology with self-aligned gates, the basis of all modern CMOS computer chips. The technology was developed by Italian physicist Federico Faggin in 1968, who later joined Intel in order to develop the very first single-chip Central Processing Unit (CPU) (Intel 4004), for which he received the National Medal of Technology and Innovation in 2010.

Advances in Integrated Circuits

ICs have consistently migrated to smaller feature sizes over the years, allowing more circuitry to be packed on each chip. This increased capacity per unit area can be used to decrease cost or increase functionality—see Moore's law which, in its modern interpretation, states that the number of transistors in an integrated circuit doubles every two years. In general, as the feature size shrinks, almost everything improves—the cost per transistor and the switching power consumption per transistor go down, and the speed goes up—see Dennard scaling. However, ICs with nanometer-scale devices are not without their problems, principal among which is leakage current, although innovations in high-κ dielectrics aim to solve these problems. Since these speed and power consumption gains are apparent to the end user, there is fierce competition among the manufacturers to use finer geometries. This process, and the expected progress over the next few years, was described for many years by the International Technology Roadmap for Semiconductors (ITRS). The final ITRS was issued in 2016, and it is being replaced by the International Roadmap for Devices and Systems or IRDS.

Initially, ICs were strictly electronic devices. The success of ICs has led to the integration of other technologies, in the attempt to obtain the same advantages of small size and low cost. These technologies include mechanical devices, optics, and sensors.

The techniques perfected by the integrated circuits industry over the last three decades have been used to create very small mechanical devices driven by electricity using a technology known as microelectromechanical systems. These devices are used in a variety of commercial and military applications. Example commercial applications include DLP projectors, inkjet printers, and accelerometers and MEMS gyroscopes used to deploy automobile airbags.

Since the early 2000s, the integration of optical functionality (optical computing) into silicon chips has been actively pursued in both academic research and in industry resulting in the successful commercialization of silicon based integrated optical transceivers combining optical devices (modulators, detectors, routing) with CMOS based electronics. Integrated optical circuits are also being developed.

In current research projects, integrated circuits are also being developed for sensor applications in medical implants or other bioelectronic devices. Special sealing techniques have to be applied in such biogenic environments to avoid corrosion or biodegradation of the exposed semiconductor materials. As one of the few materials well established in CMOS technology, titanium nitride (TiN) turned out as exceptionally stable and well suited for electrode applications in medical implants.

As of 2016, the vast majority of all transistors are fabricated in a single layer on one side of a chip of silicon in a flat 2-dimensional planar process. Researchers have produced prototypes of several promising alternatives, such as:

- various approaches to stacking several layers of transistors to make a three-dimensional integrated circuit, such as through-silicon via, "monolithic 3D", stacked wire bonding, etc.

- transistors built from other materials: graphene transistors, molybdenite transistors, carbon nanotube field-effect transistor, gallium nitride transistor, transistor-like nanowire electronic devices, organic field-effect transistor, etc.

- fabricating transistors over the entire surface of a small sphere of silicon.

- modifications to the substrate, typically to make "flexible transistors" for a flexible display or other flexible electronics, possibly leading to a roll-away computer.

Computer Assisted Design

Classification

Integrated circuits can be classified into analog, digital and mixed signal (both analog and digital on the same chip).

A CMOS 4511 IC in a DIP

Digital integrated circuits can contain anywhere from one to billions of logic gates, flip-flops, multiplexers, and other circuits in a few square millimeters. The small size of these circuits allows high speed, low power dissipation, and reduced manufacturing cost compared with board-level integration. These digital ICs, typically microprocessors, DSPs, and microcontrollers, work using binary mathematics to process "one" and "zero" signals.

Among the most advanced integrated circuits are the microprocessors or "cores", which control everything from computers and cellular phones to digital microwave ovens. Digital memory chips and application-specific integrated circuits (ASICs) are examples of other families of integrated circuits that are important to the modern information society. While the cost of designing and developing a complex integrated circuit is quite high, when spread across typically millions of production units the individual IC cost is minimized. The performance of ICs is high because the small size allows short traces which in turn allows low power logic (such as CMOS) to be used at fast switching speeds.

In the 1980s, programmable logic devices were developed. These devices contain circuits whose logical function and connectivity can be programmed by the user, rather than being fixed by the integrated circuit manufacturer. This allows a single chip to be programmed to implement different LSI-type functions such as logic gates, adders and registers. Current devices called field-programmable gate arrays (FPGAs) can (as of 2016) implement the equivalent of millions of gates in parallel and operate up to 1 GHz.

Analog ICs, such as sensors, power management circuits, and operational amplifiers, work by processing continuous signals. They perform functions like amplification, active filtering, demodulation, and mixing. Analog ICs ease the burden on circuit designers by having expertly designed analog circuits available instead of designing a difficult analog circuit from scratch.

ICs can also combine analog and digital circuits on a single chip to create functions such as A/D converters and D/A converters. Such mixed-signal circuits offer smaller size and lower cost, but must carefully account for signal interference. Prior to the late 1990s, radios could not be fabricated in the same low-cost CMOS processes as microprocessors. But since 1998, a large number of radio chips have been developed using CMOS processes. Examples include Intel's DECT cordless phone, or 802.11 (Wi-Fi) chips created by Atheros and other companies.

Modern electronic component distributors often further sub-categorize the huge variety of integrated circuits now available:

- Digital ICs are further sub-categorized as logic ICs, memory chips, interface ICs (level shifters, serializer/deserializer, etc.), Power Management ICs, and programmable devices.

- Analog ICs are further sub-categorized as linear ICs and RF ICs.

- mixed-signal integrated circuits are further sub-categorized as data acquisition ICs (including A/D converters, D/A converter, digital potentiometers) and clock/timing ICs.

Manufacturing

Fabrication

Rendering of a small standard cell with three metal layers (dielectric has been removed).
The sand-colored structures are metal interconnect, with the vertical pillars being contacts, typically plugs
of tungsten. The reddish structures are polysilicon gates, and the solid at the bottom is the crystalline silicon bulk.

The semiconductors of the periodic table of the chemical elements were identified as the most likely materials for a *solid-state vacuum tube*. Starting with copper oxide, proceeding to germanium, then silicon, the materials were systematically studied in the 1940s and 1950s. Today, monocrystalline silicon is the main substrate used for ICs although some III-V compounds of the periodic table such as gallium arsenide are used for specialized applications like LEDs, lasers, solar cells and the highest-speed integrated circuits. It took decades to perfect methods of creating crystals without defects in the crystalline structure of the semiconducting material.

Semiconductor ICs are fabricated in a planar process which includes three key process steps – imaging, deposition and etching. The main process steps are supplemented by doping and cleaning.

Schematic structure of a CMOS chip, as built in the early 2000s. The graphic shows LDD-MISFET's on an SOI substrate with five metallization layers and solder bump for flip-chip bonding. It also shows the section for FEOL (front-end of line), BEOL (back-end of line) and first parts of back-end process.

Mono-crystal silicon wafers (or for special applications, silicon on sapphire or gallium arsenide wafers) are used as the *substrate*. Photolithography is used to mark different areas of the substrate to be doped or to have polysilicon, insulators or metal (typically aluminium) tracks deposited on them.

- Integrated circuits are composed of many overlapping layers, each defined by photolithography, and normally shown in different colors. Some layers mark where various dopants are diffused into the substrate (called diffusion layers), some define where additional ions are implanted (implant layers), some define the conductors (polysilicon or metal layers), and some define the connections between the conducting layers (via or contact layers). All components are constructed from a specific combination of these layers.

- In a self-aligned CMOS process, a transistor is formed wherever the gate layer (polysilicon or metal) crosses a diffusion layer.

- Capacitive structures, in form very much like the parallel conducting plates of a traditional electrical capacitor, are formed according to the area of the "plates", with insulating material between the plates. Capacitors of a wide range of sizes are common on ICs.

- Meandering stripes of varying lengths are sometimes used to form on-chip resistors, though most logic circuits do not need any resistors. The ratio of the length of the resistive structure to its width, combined with its sheet resistivity, determines the resistance.

- More rarely, inductive structures can be built as tiny on-chip coils, or simulated by gyrators.

Since a CMOS device only draws current on the *transition* between logic states, CMOS devices consume much less current than bipolar devices.

A random-access memory is the most regular type of integrated circuit; the highest density devices are thus memories; but even a microprocessor will have memory on the chip. Although the structures are intricate – with widths which have been shrinking for decades – the layers remain much thinner than the device widths. The layers of material are fabricated much like a photographic process, although light waves in the visible spectrum cannot be used to "expose" a layer of material, as they would be too large for the features. Thus photons of higher frequencies (typically ultraviolet) are used to create the patterns for each layer. Because each feature is so small, electron microscopes are essential tools for a process engineer who might be debugging a fabrication process.

Each device is tested before packaging using automated test equipment (ATE), in a process known as wafer testing, or wafer probing. The wafer is then cut into rectangular blocks, each of which is called a *die*. Each good die (plural *dice*, *dies*, or *die*) is then connected into a package using aluminium (or gold) bond wires which are thermosonically bonded to *pads*, usually found around the edge of the die. . Thermosonic bonding was first introduced by A. Coucoulas which provided a reliable means of forming these vital electrical connections to the outside world. After packaging, the devices go through final testing on the same or similar ATE used during wafer probing. Industrial CT scanning can also be used. Test cost can account for over 25% of the cost of fabrication on lower-cost products, but can be negligible on low-yielding, larger, or higher-cost devices.

As of 2005, a fabrication facility (commonly known as a *semiconductor fab*) costs over US$1 billion to construct. The cost of a fabrication facility rises over time (Rock's law) because much of the operation is automated. Today, the most advanced processes employ the following techniques:

- The wafers are up to 300 mm in diameter (wider than a common dinner plate).

- Use of 32 nanometer or smaller chip manufacturing process. Intel, IBM, NEC, and AMD are using ~32 nanometers for their CPU chips. IBM and AMD introduced immersion lithography for their 45 nm processes

- Copper interconnects where copper wiring replaces aluminium for interconnects.

- Low-K dielectric insulators.

- Silicon on insulator (SOI).

- Strained silicon in a process used by IBM known as strained silicon directly on insulator (SSDOI).

- Multigate devices such as tri-gate transistors being manufactured by Intel from 2011 in their 22 nm process.

Packaging

The earliest integrated circuits were packaged in ceramic flat packs, which continued to be used by the military for their reliability and small size for many years. Commercial circuit packaging quickly moved to the dual in-line package (DIP), first in ceramic and later in plastic. In the 1980s pin counts of VLSI circuits exceeded the practical limit for DIP packaging, leading to pin grid array (PGA) and leadless chip carrier (LCC) packages. Surface mount packaging appeared in the early 1980s and became popular in the late 1980s, using finer lead pitch with leads formed as either gull-wing or J-lead, as exemplified by the small-outline integrated circuit (SOIC) package – a carrier which occupies an area about 30–50% less than an equivalent DIP and is typically 70% thinner. This package has "gull wing" leads protruding from the two long sides and a lead spacing of 0.050 inches.

A Soviet MSI nMOS chip made in 1977, part of a four-chip calculator set designed in 1970

In the late 1990s, plastic quad flat pack (PQFP) and thin small-outline package (TSOP) packages became the most common for high pin count devices, though PGA packages are still often used for high-end microprocessors. Intel and AMD are currently transitioning from PGA packages on high-end microprocessors to land grid array (LGA) packages.

Ball grid array (BGA) packages have existed since the 1970s. Flip-chip Ball Grid Array packages, which allow for much higher pin count than other package types, were developed in the 1990s. In an FCBGA package the die is mounted upside-down (flipped) and connects to the package balls via a package substrate that is similar to a printed-circuit board rather than by wires. FCBGA packages allow an array of input-output signals (called Area-I/O) to be distributed over the entire die rather than being confined to the die periphery.

Traces going out of the die, through the package, and into the printed circuit board have very different electrical properties, compared to on-chip signals. They require special design techniques and need much more electric power than signals confined to the chip itself.

When multiple dies are put in one package, the result is a System in Package, or SiP. A Multi-Chip Module, or MCM, is created by combining multiple dies on a small substrate often made of ceramic. The distinction between a big MCM and a small printed circuit board is sometimes fuzzy.

Chip Labeling and Manufacture Date

Most integrated circuits are large enough to include identifying information. Four common sections are the manufacturer's name or logo, the part number, a part production batch number and serial number, and a four-digit date-code to identify when the chip was manufactured. Extremely small surface mount technology parts often bear only a number used in a manufacturer's lookup table to find the chip characteristics.

The manufacturing date is commonly represented as a two-digit year followed by a two-digit week code, such that a part bearing the code 8341 was manufactured in week 41 of 1983, or approximately in October 1983.

Intellectual Property

The possibility of copying by photographing each layer of an integrated circuit and preparing photomasks for its production on the basis of the photographs obtained is a reason for the introduction of legislation for the protection of layout-designs. The Semiconductor Chip Protection Act of 1984 established intellectual property protection for photomasks used to produce integrated circuits.

A diplomatic conference was held at Washington, D.C., in 1989, which adopted a Treaty on Intellectual Property in Respect of Integrated Circuits (IPIC Treaty).

The Treaty on Intellectual Property in respect of Integrated Circuits, also called Washington Treaty or IPIC Treaty (signed at Washington on 26 May 1989) is currently not in force, but was partially integrated into the TRIPS agreement.

National laws protecting IC layout designs have been adopted in a number of countries.

Other Developments

Future developments seem to follow the multi-core multi-microprocessor paradigm, already used by Intel and AMD multi-core processors. Rapport Inc. and IBM started shipping the KC256 in 2006, a 256-core microprocessor. Intel, as recently as February–August 2011, unveiled a prototype, "not for commercial sale" chip that bears 80 cores. Each core is capable of handling its own task independently of the others. This is in response to the heat-versus-speed limit that is about to be reached using existing transistor technology. This design provides a new challenge to chip programming. Parallel programming languages such as the open-source X10 programming language are designed to assist with this task.

Generations of Ics

In the early days of simple integrated circuits, the technology's large scale limited each chip to only a few transistors, and the low degree of integration meant the design process was relatively simple. Manufacturing yields were also quite low by today's standards. As the technology progressed, millions, then billions of transistors could be placed on one chip, and good designs required thorough planning, giving rise to the field of Electronic Design Automation, or EDA.

Name	Signification	Year	Transistors number	Logic gates number
SSI	*small-scale integration*	1964	1 to 10	1 to 12
MSI	*medium-scale integration*	1968	10 to 500	13 to 99

LSI	large-scale integration	1971	500 to 20,000	100 to 9,999
VLSI	very large-scale integration	1980	20,000 to 1,000,000	10,000 to 99,999
ULSI	ultra-large-scale integration	1984	1,000,000 and more	100,000 and more

SSI, MSI and LSI

The first integrated circuits contained only a few transistors. Early digital circuits containing tens of transistors provided a few logic gates, and early linear ICs such as the Plessey SL201 or the Philips TAA320 had as few as two transistors. The number of transistors in an integrated circuit has increased dramatically since then. The term "large scale integration" (LSI) was first used by IBM scientist Rolf Landauer when describing the theoretical concept; that term gave rise to the terms "small-scale integration" (SSI), "medium-scale integration" (MSI), "very-large-scale integration" (VLSI), and "ultra-large-scale integration" (ULSI). The early integrated circuits were SSI.

SSI circuits were crucial to early aerospace projects, and aerospace projects helped inspire development of the technology. Both the Minuteman missile and Apollo program needed lightweight digital computers for their inertial guidance systems. Although the Apollo guidance computer led and motivated integrated-circuit technology, it was the Minuteman missile that forced it into mass-production. The Minuteman missile program and various other Navy programs accounted for the total $4 million integrated circuit market in 1962, and by 1968, U.S. Government space and defense spending still accounted for 37% of the $312 million total production. The demand by the U.S. Government supported the nascent integrated circuit market until costs fell enough to allow firms to penetrate the industrial, and eventually, the consumer markets. The average price per integrated circuit dropped from $50.00 in 1962 to $2.33 in 1968. Integrated circuits began to appear in consumer products by the turn of the decade, a typical application being FM inter-carrier sound processing in television receivers.

The first MOS chips were small-scale integration chips for NASA satellites.

The next step in the development of integrated circuits, taken in the late 1960s, introduced devices which contained hundreds of transistors on each chip, called "medium-scale integration" (MSI).

In 1964, Frank Wanlass demonstrated a single-chip 16-bit shift register he designed, with an incredible (at the time) 120 transistors on a single chip.

MSI devices were attractive economically because while they cost little more to produce than SSI devices, they allowed more complex systems to be produced using smaller circuit boards, less assembly work (because of fewer separate components), and a number of other advantages.

Further development, driven by the same economic factors, led to "large-scale integration" (LSI) in the mid-1970s, with tens of thousands of transistors per chip.

The masks used to process and manufacture SSI, MSI and early LSI and VLSI devices (such as the microprocessors of the early 1970s) were mostly created by hand, often using Rubylith-tape or similar. For large or complex ICs (such as memories or processors), this was often done by specially hired layout people under supervision of a team of engineers, who would also, along with the circuit designers, inspect and verify the correctness and completeness of each mask. However, modern VLSI devices contain so many transistors, layers, interconnections, and other features that it is no longer feasible to check the masks or do the original design by hand. The engineer depends on computer programs and other hardware aids to do most of this work.

Integrated circuits such as 1K-bit RAMs, calculator chips, and the first microprocessors, that began to be manufactured in moderate quantities in the early 1970s, had under 4000 transistors. True LSI circuits, approaching 10,000 transistors, began to be produced around 1974, for computer main memories and second-generation microprocessors.

Some SSI and MSI chips, like discrete transistors, are still mass-produced, both to maintain old equipment and build new devices that require only a few gates. The 7400 series of TTL chips, for example, has become a de facto standard and remains in production.

VLSI

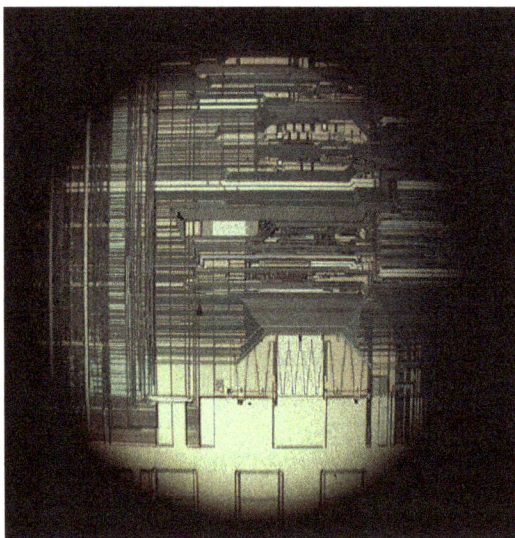

Upper interconnect layers on an Intel 80486DX2 microprocessor die

The final step in the development process, starting in the 1980s and continuing through the present, was "very-large-scale integration" (VLSI). The development started with hundreds of thousands of transistors in the early 1980s, and continues beyond ten billion transistors as of 2016.

Multiple developments were required to achieve this increased density. Manufacturers moved to smaller design rules and cleaner fabrication facilities, so that they could make chips with more transistors and maintain adequate yield. The path of process improvements was summarized by the International Technology Roadmap for Semiconductors (ITRS). Design tools improved enough to make it practical to finish these designs in a reasonable time. The more energy-efficient CMOS replaced NMOS and PMOS, avoiding a prohibitive increase in power consumption.

In 1986 the first one-megabit RAM chips were introduced, containing more than one million transistors. Microprocessor chips passed the million-transistor mark in 1989 and the billion-transistor mark in 2005. The trend continues largely unabated, with chips introduced in 2007 containing tens of billions of memory transistors.

ULSI, WSI, SOC and 3D-IC

To reflect further growth of the complexity, the term *ULSI* that stands for "ultra-large-scale integration" was proposed for chips of more than 1 million transistors.

Wafer-scale integration (WSI) is a means of building very large integrated circuits that uses an entire silicon wafer to produce a single "super-chip". Through a combination of large size and reduced packaging, WSI could lead to dramatically reduced costs for some systems, notably massively parallel supercomputers. The name is taken from the term Very-Large-Scale Integration, the current state of the art when WSI was being developed.

A system-on-a-chip (SoC or SOC) is an integrated circuit in which all the components needed for a computer or other system are included on a single chip. The design of such a device can be complex and costly, and building disparate components on a single piece of silicon may compromise the efficiency of some elements. However, these drawbacks are offset by lower manufacturing and assembly costs and by a greatly reduced power budget: because signals among the components are kept on-die, much less power is required.

A three-dimensional integrated circuit (3D-IC) has two or more layers of active electronic components that are integrated both vertically and horizontally into a single circuit. Communication between layers uses on-die signaling, so power consumption is much lower than in equivalent separate circuits. Judicious use of short vertical wires can substantially reduce overall wire length for faster operation.

Silicon Labelling and Graffiti

To allow identification during production most silicon chips will have a serial number in one corner. It is also common to add the manufacturer's logo. Ever since ICs were created, some chip designers have used the silicon surface area for surreptitious, non-functional images or words. These are sometimes referred to as chip art, silicon art, silicon graffiti or silicon doodling.

ICs and IC Families

- The 555 timer IC

- The 741 operational amplifier

- 7400 series TTL logic building blocks

- 4000 series, the CMOS counterpart to the 7400 series

- Intel 4004, the world's first microprocessor, which led to the famous 8080 CPU and then the IBM PC's 8088, 80286, 486 etc.

- The MOS Technology 6502 and Zilog Z80 microprocessors, used in many home computers of the early 1980s

- The Motorola 6800 series of computer-related chips, leading to the 68000 and 88000 series (used in some Apple computers and in the 1980s Commodore Amiga series).

- The LM-series of analog integrated circuits.

Invention of the Integrated Circuit

The idea of integrating electronic circuits into a single device was born when the German physicist and engineer Werner Jacobi (de) developed and patented the first known integrated transistor amplifier in 1949 and the British radio engineer Geoffrey Dummer proposed to integrate a variety of standard electronic components in a monolithic semiconductor crystal in 1952. A year later, Harwick Johnson filed a patent for a prototype integrated circuit (IC).

These ideas could not be implemented by the industry in the early 1950s, but a breakthrough came in late 1958. Three people from three U.S. companies solved three fundamental problems that hindered the production of integrated circuits. Jack Kilby of Texas Instruments patented the principle of integration, created the first prototype ICs and commercialized them. Kurt Lehovec of Sprague Electric Company invented a way to electrically isolate components on a semiconductor crystal. Robert Noyce of Fairchild Semiconductor invented a way to connect the IC components (aluminium metallization) and proposed an improved version of insulation based on the planar technology by Jean Hoerni. On September 27, 1960, using the ideas of Noyce and Hoerni, a group of Jay Last's at Fairchild Semiconductor created the first operational semiconductor IC. Texas Instruments, which held the patent for Kilby's invention, started a patent war, which was settled in 1966 by the agreement on cross-licensing.

There is no consensus on who invented the IC. The American press of the 1960s named four people: Kilby, Lehovec, Noyce and Hoerni; in the 1970s the list was shortened to Kilby and Noyce, and then to Kilby, who was awarded the 2000 Nobel Prize in Physics "for his part in the invention of the integrated circuit". In the 2000s, historians Leslie Berlin, Bo Lojek and Arjun Saxena reinstated the idea of multiple IC inventors and revised the contribution of Kilby.

Prerequisites

Waiting for a Breakthrough

During and immediately after World War II a phenomenon named "the tyranny of numbers" was noticed, that is, some computational devices reached a level of complexity at which the losses from failures and downtime exceeded the expected benefits. Each Boeing B-29 (put into service in 1944) carried 300–1000 vacuum tubes and tens of thousands of passive components. The number of vacuum tubes reached thousands in advanced computers and more than 17,000 in the ENIAC (1946). Each additional component reduced the reliability of a device and lengthened the troubleshooting time. Traditional electronics reached a deadlock and a further development of electronic devices required reducing the number of their components.

Replacing a bad tube meant checking among ENIAC's 19,000 possibilities.

Changing vacuum tubes in the computer ENIAC. By the 1940s, some computational devices reached
the level at which the losses from failures and downtime outweighed the economic benefits.

The invention of the transistor in 1948 led to the expectation of a new technological revolution.
Fiction writers and journalists heralded the imminent appearance of "intelligent machines" and
robotization of all aspects of life. Although transistors did reduce the size and power consumption, they could not solve the problem of reliability of complex electronic devices. On the contrary,
dense packing of components in small devices hindered their repair. While the reliability of discrete components was brought to the theoretical limit in the 1950s, there was no improvement in
the connections between the components.

Idea of Integration

Early developments of the integrated circuit go back to 1949, when the German engineer Werner
Jacobi (de) (Siemens AG) filed a patent for an integrated-circuit-like semiconductor amplifying
device showing five transistors on a common substrate in a 3-stage amplifier arrangement with
two transistors working „upside-down" as impedance converter. Jacobi disclosed small and cheap
hearing aids as typical industrial applications of his patent. An immediate commercial use of his
patent has not been reported.

On May 7, 1952, the British radio engineer Geoffrey Dummer formulated the idea of integration in
a public speech in Washington:

With the advent of the transistor and the work in semiconductors generally, it seems now to be
possible to envisage electronic equipment in a solid block with no connecting wires. The block may
consist of layers of insulating, conducting, rectifying and amplifying materials, the electrical functions being connected by cutting out areas of the various layers.

Johnson's integrated generator (1953; variants with lumped and distributed capacitances).
Inductances L, load resistor Rk and sources Бк и Бб are external.

Dummer later became famous as "the prophet of integrated circuits", but not as their inventor. In
1956 he produced an IC prototype by growth from the melt, but his work was deemed impractical
by the UK Ministry of Defence, because of the high cost and inferior parameters of the IC compared to discrete devices.

In May 1952, Sidney Darlington filed a patent application in the United States for a structure with two or three transistors integrated onto a single chip in various configurations; in October 1952, Bernard Oliver filed a patent application for a method of manufacturing three electrically connected planar transistors on one semiconductor crystal. On May 21, 1953, Harwick Johnson filed a patent application for a method of forming various electronic components – transistors, resistors, lumped and distributed capacitances – on a single chip. Johnson described three ways of producing an integrated one-transistor oscillator. All of them used a narrow strip of a semiconductor with a bipolar transistor on one end and differed in the methods of producing the transistor. The strip acted as a series of resistors; the lumped capacitors were formed by fusion whereas inverse-biased p-n junctions acted as distributed capacitors. Johnson did not offer a technological procedure, and it is not known whether he produced an actual device. In 1959, a variant of his proposal was implemented and patented by Jack Kilby.

Functional Electronics

The leading US electronics companies (Bell Labs, IBM, RCA and General Electric) sought solution to "the tyranny of numbers" in the development of discrete components that implemented a given function with a minimum number of attached passive elements. During the vacuum tube era, this approach allowed to reduce the cost of a circuit at the expense of its operation frequency. For example, a memory cell of the 1940s consisted of two triodes and a dozen passive components and ran at frequencies up to 200 kHz. A MHz response could be achieved with two pentodes and six diodes per cell. This cell could be replaced by one thyratron with a load resistor and an input capacitor, but the operating frequency of such circuit did not exceed a few kHz.

In 1952, Jewell James Ebers from Bell Labs developed a prototype solid-state analog of thyratron – a four-layer transistor, or thyristor. William Shockley simplified its design to a two-terminal "four-layer diode" (Shockley diode) and attempted its industrial production. Shockley hoped that the new device will replace the polarized relay in telephone exchange; however, the reliability of Shockley diodes was unacceptably low, and his company went into decline.

At the same time, works on thyristor circuits were carried at Bell Labs, IBM and RCA. Ian Munro Ross and David D'Azaro (Bell Labs) experimented with thyristor-based memory cells. Joe Logue and Rick Dill (IBM) were building counters using monojunction transistors. Torkel Wallmark and Harwick Johnson (RCA) used both the thyristors and field-effect transistors. The works of 1955–1958 that used germanium thyristors were fruitless. Only in the summer of 1959, after the inventions of Kilby, Lehovec and Hoerni became publicly known, D'Azaro reported an operational shift register based on silicon thyristors. In this register, one crystal containing four thyristors replaced eight transistors, 26 diodes and 27 resistors. The area of each thyristor ranged from 0.2 to 0.4 mm², with a thickness of about 0.1 mm. The circuit elements were isolated by etching deep grooves.

From the point of view of supporters of functional electronics, semiconductor era, their approach was allowed to circumvent the fundamental problems of semiconductor technology. The failures of Shockley, Ross and Wallmark proved the fallacy of this approach: the mass production of functional devices was hindered by technological barriers.

Silicon Technology

Comparison of the mesa (left) and planar (Hoerni, right) technologies. Dimensions are shown schematically.

Early transistors were made of germanium. By the mid-1950s it was replaced by silicon that allowed operation at higher temperatures. In 1954, Gordon Kidd Teal from Texas Instruments produced the first silicon transistor, which became commercial in 1955. Also in 1954, Fuller and Dittsenberger published a fundamental study of diffusion in silicon, and Shockley suggested using this technology to form p-n junctions with a given profile of the impurity concentration.

In early 1955, Carl Frosch from Bell Labs developed wet oxidation of silicon, and in the next two years, Frosch, Moll, Fuller and Holonyak brought it to the mass production. This accidental discovery revealed the second fundamental advantage of silicon over germanium: contrary to germanium oxides, "wet" silica is a physically strong and chemically inert electrical insulator.

On December 1, 1957, Jean Hoerni first proposed a planar technology of bipolar transistors. In this process, all the p-n junctions were covered by a protective layer, which should significantly improve reliability. However, in 1957, this proposal was considered technically impossible. The formation of the emitter of an n-p-n transistor required diffusion of phosphorus, and the work of Frosch suggested that SiO_2 does not block such diffusion. In March 1959, Chi-Tang Sah, a former colleague of Hoerni, pointed Hoerni and Noyce to an error in the conclusions of Frosch. Frosch used too thin oxide layers, whereas the experiments of 1957–1958 showed that a thick layer of oxide can stop the phosphorus diffusion. Armed with this knowledge, by March 12, 1959 Hoerni made the first prototype of a planar transistor, and on May 1, 1959 filed a patent application for the invention of the planar process. In April 1960, Fairchild launched the planar transistor 2N1613, and by October 1960 completely abandoned the mesa transistor technology. By the mid-1960s, the planar process has become the main technology of producing transistors and monolithic integrated circuits.

Three Problems of Microelectronics

The creation of the integrated circuit was hindered by three fundamental problems, which were formulated by Wallmark in 1958:

1. Integration. In 1958, there was no way of forming many different electronic components in one semiconductor crystal. Alloying was not suited to the IC and the latest mesa technology had serious problems with reliability.

2. Isolation. There was no technology to electrically isolate components on one semiconductor crystal.

3. Connection. There was no effective way to create electrical connections between the components of an IC, except for the extremely expensive and time-consuming connection using gold wires.

It happened so that three different companies held the key patents to each of these problems. Sprague Electric Company decided not to develop ICs, Texas Instruments limited itself to an incomplete set of technologies, and only Fairchild Semiconductor combined all the techniques required for a commercial production of monolithic ICs.

Integration by Jack Kilby

Kilby's Invention

In May 1958, Jack Kilby, an experienced radio engineer and a veteran of World War II, started working at Texas Instruments. At first, he had no specific tasks and had to find himself a suitable topic in the general direction of "miniaturization". He had a chance of either finding a radically new research direction or blend into a multimillion-dollar project on the production of military circuits. In the summer of 1958, Kilby formulated three features of integration:

1. The only thing that a semiconductor company can successfully produce is semiconductors.

2. All circuit elements, including resistors and capacitors can be made of a semiconductor.

3. All circuit components can be formed on one semiconductor crystal, adding only the interconnections.

Comparison of the oscillators by Johnson (with an alloyed transistor, length: 10 mm, width: 1.6 mm) and Kilby (with mesa transistor).

On August 28, 1958, Kilby assembled the first prototype of an IC using discrete components and received approval for implementing it on one chip. He had access to technologies that allowed to form mesa transistors, mesa diodes and capacitors based on p-n junctions on a germanium (but not silicon) chip, and the bulk material of the chip could be used for resistors. The standard Texas Instruments chip for the production of 25 (5×5) mesa transistors was 10×10 mm in size. Kilby cut it into five-transistor 10×1.6 mm strips, but later used not more than two of them. On September 12, he presented the first IC prototype, which was a single-transistor oscillator with a distributed RC feedback, repeating the idea and the circuit in the 1953 patent by Johnson. On September 19, he made the second prototype, a two-transistor trigger. He described these ICs, referencing the Johnson's patent, in his U.S. Patent 3,138,743.

Between February and May 1959 Kilby filed a series of applications: U.S. Patent 3,072,832, U.S. Patent 3,138,743, U.S. Patent 3,138,744, U.S. Patent 3,115,581 and U.S. Patent 3,261,081 (their carry very different numbers because of the spread in the issue dates − the larger the number the later the issue). According to Arjun Saxena, the application date for the key patent 3,138,743 is uncertain: while the patent and the book by Kilby set it to February 6, 1959, it could not be confirmed by the application archives of the federal patent office. He suggested that the initial application was filed on February 6 and lost, and the (preserved) resubmission was received by the patent office on 6 May 1959 − the same date as the applications for the patents 3,072,832 and 3,138,744. Texas Instruments introduced the inventions by Kilby to the public on March 6, 1959.

None of these patents solved the problem of isolation and interconnection − the components were separated by cutting grooves on the chip and connected by gold wires. Thus these ICs were of the hybrid rather than monolithic type. However, Kilby demonstrated that various circuit elements: active components, resistors, capacitors and even small inductances can be formed on one chip.

Commercialization Attempts

In autumn 1958, Texas Instruments introduced the yet non-patented idea of Kilby to military customers. While most divisions rejected it as unfit to the existing concepts, the US Air Force decided that this technology complies with their molecular electronics program, and ordered production of prototype ICs, which Kilby named "functional electronic blocks". Westinghouse added epitaxy to the Texas Instruments technology and received a separate order from the US military in January 1960.

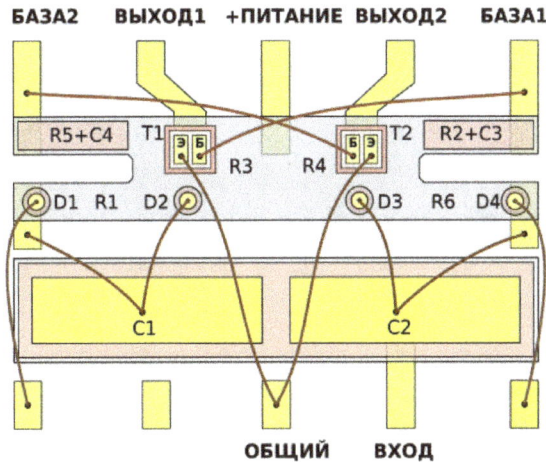

Topology of the double-crystal multivibrator IC TI 502. The numbering corresponds to File:TI 502 schematic.png. Each crystal is 5 mm long. Proportions are slightly altered for presentation purposes.

In October 1961, Texas Instruments built for the Air Force a demonstration "molecular computer" with a 300-bit memory based on the #587 ICs of Kilby. Harvey Kreygon packed this computer into a volume of a little over 100 cm³. In December 1961, the Air Force accepted the first analog device created within the molecular electronics program – a radio receiver. It uses costly ICs, which had less than 10–12 components and a high percentage of failed devices. This generated an opinion that ICs can only justify themselves for aerospace applications. However, the aerospace industry rejected those ICs for the low radiation hardness of their mesa transistors.

In April 1960, Texas Instruments announced multivibrator #502 as the world's first integrated circuit available on the market. The company assured that contrary to the competitors they actually sell their product, at a price of US$450 per unit or US$300 for quantities larger than 100 units. However, the sales began only in the summer of 1961, and the price was higher than announced. The #502 schematic contained two transistors, four diodes, six resistors and two capacitors, and repeated the traditional discrete circuitry. The device contained two Si strips of 5 mm length inside a metal-ceramic housing. One strip contained input capacitors; the other accommodated mesa transistors and diodes, and its grooved body was used as six resistors. Gold wires acted as interconnections.

Isolation by p-n Junction

Solution by Kurt Lehovec

In late 1958, Kurt Lehovec, a scientist working at the Sprague Electric Company, attended a seminar at Princeton where Wallmark outlined his vision of the fundamental problems in microelectronics. On his way back to Massachusetts, Lehovec found a simple solution to the isolation problem which used the p-n junction:

It is well-known that a p-n junction has a high impedance to electric current, particularly if biased in the so-called blocking direction, or with no bias applied. Therefore, any desired degree of electrical insulation between two components assembled on the same slice can be achieved by having a sufficiently large number of p-n junctions in series between two semiconducting regions on which said components are assembled. For most circuits, one to three junctions will be sufficient.

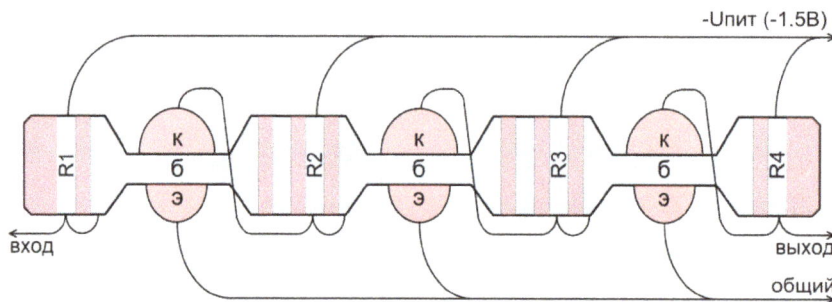

Cross-section of a three-stage amplifier (three transistors, four resistors) from U.S. Patent 3,029,366. Blue areas: n-type conductivity, red: p-type, length: 2.2 mm, thickness: 0.1 mm.

Lehovec tested his idea using the technologies of making transistors that were available at Sprague. His device was a linear structure 2.2×0.5×0.1 mm in size, which was divided into isolated n-type cells (bases of the future transistors) by p-n junctions. Layers and transitions were formed by growth from the melt. The conductivity type was determined by the pulling speed of the crystal: an indium-rich p-type layer was formed at a slow speed, whereas an arsenic-rich n-type layer was produced at a high speed. The collectors and emitters of the transistors were created by welding indium beads. All electrical connections were made by hand, using gold wires.

The management of Sprague showed no interest to the invention by Lehovec. Nevertheless, on April 22, 1959 he filed a patent application at his own expense, and then left the United States for two years. Because of this disengagement, Gordon Moore concluded that Lehovec should not be considered as an inventor of the integrated circuit.

Solution by Robert Noyce

On January 14, 1959, Jean Hoerni introduced his latest version of the planar process to Robert Noyce and a patent attorney John Rallza at Fairchild Semiconductor. A memo of this event by Hoerni was the basis of a patent application for the invention of a planar process, filed in May 1959, and implemented in U.S. Patent 3,025,589 (the planar process) and U.S. Patent 3,064,167 (the planar transistor). On January 20, 1959, Fairchild managers met with Edward Keonjian, the developer of the onboard computer for the rocket "Atlas", to discuss the joint development of hybrid digital ICs for his computer. These events probably led Robert Noyce to return to the idea of integration.

On January 23, 1959, Noyce documented his vision of the planar integrated circuit, essentially re-inventing the ideas of Kilby and Lehovec on the base of the Hoerni's planar process. Noyce claimed in 1976 that in January 1959 he did not know about the work of Lehovec.

As an example, Noyce described an integrator that he discussed with Keonjian. Transistors, diodes and resistors of that hypothetical device were isolated from each other by p-n junctions, but in a different manner from the solution by Lehovec. Noyce considered the IC manufacturing process as follows. It should start with a chip of highly resistive intrinsic (undoped) silicon passivated with an oxide layer. The first photolithography step aims to open windows corresponding to the planned devices, and diffuse impurities to create low-resistance "wells" through the entire thickness of the chip. Then traditional planar devices are formed inside those wells. Contrary to the solution by Lehovec, this approach allowed creation of two-dimensional structures with a potentially unlimited number of devices on a chip.

After formulating his idea, Noyce shelved it for several months due to pressing company matters, and returned to it only by March 1959. It took him six months to prepare a patent application, which was then rejected by the US Patent Office because they already received the application by Lehovec. Noyce revised his application and in 1964 received U.S. Patent 3,150,299 and U.S. Patent 3,117,260.

Invention of Metallization

In early 1959, Noyce solved another important problem, the problem of interconnections that hindered mass-production of ICs. According to the colleagues from the traitorous eight his idea was self-evident: of course, the passivating oxide layer forms a natural barrier between the chip and the metallization layer. According to Turner Hasty, who worked with Kilby and Noyce, Noyce planned to make the microelectronic patents of Fairchild accessible to a wide range of companies, similar to Bell Labs which in 1951–1952 released their transistor technologies.

Noyce submitted his application on July 30, 1959, and on April 25, 1961 received U.S. Patent 2,981,877. According to the patent, the invention consisted of preserving the oxide layer, which separated the metallization layer from the chip (except for the contact window areas), and of depositing the metal layer so that it is firmly attached to the oxide. The deposition method was not yet known, and the proposals by Noyce included vacuum deposition of aluminium through a mask and deposition of a continuous layer, followed by photolithography and etching off the excess metal. According to Saxena, the patent by Noyce, with all its drawbacks, accurately reflects the fundamentals of the modern IC technologies.

In his patent, Kilby also mentions the use of metallization layer. However, Kilby favored thick coating layers of different metals (aluminium, copper or antimony-doped gold) and silicon monoxide instead of the dioxide. These ideas were not adopted in the production of ICs.

First Semiconductor ICs

In August 1959, Noyce formed at Fairchild a group to develop integrated circuits. On May 26, 1960, this group, led by Jay Last, produced the first planar integrated circuit. This prototype was not monolithic – two pairs of its transistors were isolated by cutting a groove on the chip, according to the patent by Last. The initial production stages repeated the Hoerni's planar process. Then the 80-micron-thick crystal was glued, face down, to the glass substrate, and additional photolithography was carried on the back surface. Deep etching created a groove down to the front surface. Then the back surface was covered with an epoxy resin, and the chip was separated from the glass substrate.

Logical NOR IC from the computer that controlled the Apollo spacecraft

In August 1960, Last started working on the second prototype, using the isolation by p-n junction proposed by Noyce. Robert Norman developed a trigger circuit on four transistors and five resistors, whereas Isy Haas and Lionel Kattner developed the process of boron diffusion to form the insulating regions. The first operational device was tested in September 27, 1960 – this was the first planar and monolithic integrated circuit.

Fairchild Semiconductor did not realize the importance of this work. Vice president of marketing believed that Last was wasting the company resources and that the project should be terminated. In January 1961, Last, Hoerni and their colleagues from the "traitorous eight" Kleiner and Roberts left Fairchild and headed Amelco. David Allison, Lionel Kattner and some other technologists left Fairchild to establish a direct competitor, the company Signetics.

Despite the departure of their leading scientists and engineers, in March 1961 Fairchild announced their first commercial IC series, named "Micrologic", and then spent a year on creating a family of logic ICs. By that time ICs were already produced by their competitors. Texas Instruments abandoned the IC designs by Kilby and received a contract for a series of planar ICs for space satellites, and then for the LGM-30 Minuteman ballistic missiles. Whereas the ICs for the onboard computers of the Apollo spacecraft were designed by Fairchild, most of them were produced by Raytheon and Philco Ford. Each of these computers contained about 5,000 standard logic ICs, and during their manufacture, the price for an IC dropped from US$1,000 to US$20–30. In this way, NASA and the Pentagon prepared the ground for the non-military IC market.

The resistor-transistor logic of first ICs by Fairchild and Texas Instruments was vulnerable to electromagnetic interference, and therefore in 1964 both companies replaced it by the diode-transistor logic . Signetics released the diode-transistor family Utilogic back in 1962, but fell behind Fairchild and Texas Instruments with the expansion of production. Fairchild was the leader in the number of ICs sold in 1961–1965, but Texas Instruments was ahead in the revenue: 32% of the IC market in 1964 compared to 18% of Fairchild.

The above logic ICs were built from standard components, with sizes and configurations defined by the technological process, and all the diodes and transistors on one IC were of the same type. The use of different transistor types was first proposed by Tom Long at Sylvania in 1961–1962. In late 1962, Sylvania launched the first family of transistor-transistor logic (TTL) ICs, which became a commercial success. Bob Widlar from Fairchild made a similar breakthrough in 1964–1965 in analog ICs (operational amplifiers).

Patent Wars of 1962–1966

In 1959–1961 years, when Texas Instruments and Westinghouse worked in parallel on aviation "molecular electronics", their competition had a friendly character. The situation changed in 1962 when Texas Instruments started to zealously pursue the real and imaginary infringers of their patents and received the nicknames "The Dallas legal firm" and "semiconductor cowboys". This example was followed by some other companies. Nevertheless, the IC industry continued to develop no matter the patent disputes.

Texas Instruments v. Westinghouse

> In 1962–1963, when these companies have adopted the planar process, the Westinghouse engineer Hung-Chang Lin invented the lateral transistor. In the usual planar process, all transistors have the same conductivity type, typically n-p-n, whereas the invention by Lin allowed to create n-p-n and p-n-p transistors on one chip. The military orders that were anticipated by Texas Instruments went to Westinghouse. TI filed a case, which was settled out of court.

Texas Instruments v. Sprague

> On April 10, 1962, Lehovec received a patent for isolation by p-n junction. Texas Instruments immediately filed a court case claiming that the isolation problem was solved in their earlier patent filed by Kilby. Robert Sprague, the founder of Sprague, considered the case hopeless and was going to give up the patent rights, was convinced otherwise by Lehovec. Four years later, Texas Instruments hosted in Dallas an arbitration hearing with demonstrations of the Kilby's inventions and depositions by experts. However, Lehovec conclusively proved that Kilby did not mention isolation of components. His priority on the isolation patent was finally acknowledged in April 1966.

Raytheon v. Fairchild

> On May 20, 1962, Jean Hoerni, who had already left Fairchild, received the first patent on the planar technology. Raytheon believed that Hoerni repeated the patent held by Jules Andrews and Raytheon and filed a court case. While appearing similar in the photolithography, diffusion and etching processes, the approach of Andrews had a fundamental flaw: it involved the complete removal of the oxide layer after each diffusion. On the contrary, in the process of Hoerni the "dirty" oxide was kept. Raytheon withdrew their claim and obtained a license from Fairchild.

Hughes v. Fairchild

Hughes Aircraft sued Fairchild arguing that their researchers developed the Hoerni's process earlier. According to Fairchild lawyers, this case was baseless, but could take a few years, during which Fairchild could not sell the license to Hoerni's process. Therefore, Fairchild chose to settle with Hughes out of court. Hughes acquired the rights to one of the seventeen points of the Hoerni's patent, and then exchanged it for a small percentage of the future licensing incomes of Fairchild.

Texas Instruments v. Fairchild

In their legal wars, Texas Instruments focused on their largest and most technologically advanced competitor, Fairchild Semiconductor. Their cases hindered not the production at Fairchild, but the sale of licenses for their technologies. By 1965, the planar technology of Fairchild became the industry standard, but the license to patents of Hoerni and Noyce was purchased by less than ten manufacturers, and there were no mechanisms to pursue unlicensed production. Similarly, the key patents of Kilby were bringing no income to Texas Instruments. In 1964, the patent arbitration awarded Texas Instruments the rights to four of the five key provisions of the contested patents, but both companies appealed the decision. The litigation could continue for years, if not the defeat of Texas Instruments in the dispute with Sprague in April 1966. Texas Instruments realized that they could not claim priority for the whole set of key IC patents, and lost interest in the patent war. In the summer of 1966, Texas Instruments and Fairchild agreed on the mutual recognition of patents and cross-licensing of key patents; in 1967 they were joined by Sprague.

Japan v. Fairchild

In the early 1960s, both Fairchild and Texas Instruments tried to set up IC production in Japan, but were opposed by the Japan Ministry of International Trade and Industry (MITI). In 1962, MITI banned Fairchild from further investments in the factory that they already purchased in Japan, and Noyce tried to enter the Japanese market through the corporation NEC. In 1963, the management of NEC pushed Fairchild to extremely advantageous for Japan licensing terms, strongly limiting the Fairchild sales in the Japanese market. Only after concluding the deal Noyce learned that the president of NEC also chaired the MITI committee that blocked the Fairchild deals.

Japan v. Texas Instruments

In 1963, despite the negative experience with NEC and Sony, Texas Instruments tried to establish their production in Japan. For two years MITI did not give a definite answer to the request, and in 1965 Texas Instruments retaliated by threatening with embargo on the import of electronic equipment that infringed their patents. This action hit Sony in 1966 and Sharp in 1967, prompting MITI to secretly look for a Japanese partner to Texas Instruments. MITI blocked the negotiations between Texas Instruments and Mitsubishi (the owner of Sharp), and persuaded Akio Morita to make a deal with Texas Instruments "for the future of Japanese industry". Despite the secret protocols that guaranteed the Americans a share in Sony the agreement of 1967–1968 was extremely disadvantageous for Texas Instruments. For almost thirty years, Japanese companies were producing

ICs without paying royalties to Texas Instruments, and only in 1989 the Japanese court acknowledged the patent rights to the invention by Kilby. As a result, in the 1990s, all of Japanese IC manufacturers had to pay for the 30 years old patent or enter into cross-licensing agreements. In 1993, Texas Instruments earned US$520 million in license fees, mostly from Japanese companies.

Historiography of the Invention

Two Inventors: Kilby and Noyce

During the patent wars of the 1960s the press and professional community in the United States recognized that the number of the IC inventors could be rather large. The book "Golden Age of Entrepreneurship" named four people: Kilby, Lehovec, Noyce and Hoerni. Sorab Ghandhi in "Theory and Practice of Microelectronics" (1968) wrote that the patents of Lehovec and Hoerni were the high point of semiconductor technology of the 1950s and opened the way for the mass production of ICs.

In October 1966, Kilby and Noyce were awarded the Ballantine Medal from the Franklin Institute "for their significant and essential contribution to the development of integrated circuits". This event initiated the idea of two inventors. The nomination of Kilby was criticized by contemporaries who did not recognize his prototypes as "real" semiconductor ICs. Even more controversial was the nomination of Noyce: the engineering community was well aware of the role of the Moore, Hoerni and other key inventors, whereas Noyce at the time of his invention was CEO of Fairchild and did not participate directly in the creation of the first IC. Noyce himself admitted, "I was trying to solve a production problem. I wasn't trying to make an integrated circuit".

According to Leslie Berlin, Noyce became the "father of the integrated circuit" because of the patent wars. Texas Instruments picked his name because of stood on the patent they challenged and thereby "appointed" him as a sole representative of all the development work at Fairchild. In turn, Fairchild mobilized all its resources to protect the company, and thus the priority of Noyce. While Kilby was personally involved in the public relation campaigns of Texas Instruments, Noyce kept away from publicity and was substituted by Gordon Moore.

By the mid-1970s, the two-inventor version became widely accepted, and the debates between Kilby and Lehovec in professional journals in 1976–1978 did not change the situation. Hoerni, Last and Lehovec were regarded as minor players; they did not represent large corporations and were not keen for public priority debates.

In scientific articles of the 1980s, the history of IC invention was often presented as follows

While at Fairchild, Noyce developed the integrated circuit. The same concept has been invented by Jack Kilby at Texas Instruments in Dallas a few months previously. In July 1959 Noyce filed a patent for his conception of the integrated circuit. Texas Instruments filed a lawsuit for patent interference against Noyce and Fairchild, and the case dragged on for some years. Today, Noyce and Kilby are usually regarded as co-inventors of the integrated circuit, although Kilby was inducted into the Inventor's Hall of Fame as the inventor. In any event, Noyce is credited with improving the integrated circuit for its many applications in the field of microelectronics.

In 1984, the two-inventor version has been further supported by Thomas Reid in "The Chip: How Two Americans Invented the Microchip and Launched a Revolution". The book was reprinted up to 2008. Robert Wright of The New York Times criticized Reid for a lengthy description of the supporting characters involved in the invention, yet the contributions of Lehovec and Last were not mentioned, and Jean Hoerni appears in the book only as a theorist who consulted Noyce.

Paul Ceruzzi in "A History of Modern Computing" (2003) also repeated the two-inventor story and stipulated that "Their invention, dubbed at first *Micrologic*, then the *Integrated Circuit* by Fairchild, was simply another step along this path" (of miniaturization demanded by the military programs of the 1950s). Referring to the prevailing in the literature opinion, he put forward the decision of Noyce to use the planar process of Hoerni, who paved the way for the mass production of ICs, but was not included in the list of IC inventors. Ceruzzi did not cover the invention of isolation of IC components.

In 2000, the Nobel Committee awarded the Nobel Prize in Physics to Kilby "for his part in the invention of the integrated circuit". Noyce died in 1990 and thus could not be nominated; when asked during his life about the prospects of the Nobel Prize he replied "They don't give Nobel Prizes for engineering or real work". Because of the confidentiality of the Nobel nomination procedure, it is not known whether other IC inventors had been considered. Saxena argued that the contribution of Kilby was pure engineering rather than basic science, and thus his nomination violated the will of Alfred Nobel.

The two-inventor version persisted through the 2010s. Its variation puts Kilby in front, and considers Noyce as an engineer who improved the Kilby's invention. Fred Kaplan in his popular book "1959: The Year Everything Changed" (2010) spends eight pages on the IC invention and assigns it to Kilby, mentioning Noyce only in a footnote and neglecting Hoerni and Last.

Revision of the Canonical Version

In the late 1990s and 2000s a series of books presented the IC invention beyond the simplified two-person story. In 1998, Michael Riordan and Lillian Hoddson described in detail the events leading to the invention of Kilby in their book "Crystal Fire: The Birth of the Information Age". However, they stopped on that invention. Leslie Berlin in her biography of Robert Noyce (2005) included the events unfolding at Fairchild and critically evaluated the contribution of Kilby. "Connecting wire precluded production and Kilby could not know that.

In 2007, Bo Lojek opposed the two-inventor version; he described the contribution of Hoerni and Last and criticized Kilby.

In 2009, Saxena described the work of Dummer, Johnson, Stewart, Kilby, Noyce, Lehovec and Hoerni. He also played down the role of Kilby and Noyce.

Integrated Circuit Development

The integrated circuit (IC) development process is complex and arduous. The high level process for developing an integrated circuit starts with defining product requirements, progresses through architectural definition, implementation, bringup and finally productization. The various phases of the integrated circuit development process are described below. Although the phases are presented here in a straightforward fashion, in reality there is iteration and these steps may occur multiple times.

Requirements

Before an architecture can be defined some high level product goals must be defined. The requirements are usually generated by a cross functional team that addresses market opportunity, customer needs, feasibility and much more. This phase should result in a product requirements document.

Architecture

The *architecture* defines the fundamental structure, goals and principles of the product. It defines high level concepts and the intrinsic value proposition of the product. Architecture teams take into account many variables and interface with many groups. People creating the architecture generally have a significant amount of experience dealing with systems in the area for which the architecture is being created. The work product of the architecture phase is an architectural specification.

Micro-architecture

The micro-architecture is a step closer to the hardware. It implements the architecture and defines specific mechanisms and structures for achieving that implementation. The result of the micro-architecture phase is a micro-architecture specification which describes the methods used to implement the architecture.

Implementation

In the implementation phase the design itself is created using the micro-architectural specification as the starting point. This involves low level definition and partitioning, writing code, entering schematics and verification. This phase ends with a design reaching tapeout.

Bringup

After a design is created, taped-out and manufactured, actual hardware, 'first silicon', is received which is taken into the lab where it goes through *bringup*. Bringup is the process of powering, testing and characterizing the design in the lab. Numerous tests are performed starting from very simple tests such as ensuring that the device will power on to much more complicated tests which try to stress the part in various ways. The result of the bringup phase is documentation of characterization data (how well the part performs to spec) and errata (unexpected behavior).

Productization

Productization is the task of taking a design from engineering into mass production manufacturing. Although a design may have successfully met the specifications of the product in the lab during the bringup phase there are many challenges that face product engineers when trying to mass-produce those designs. The IC must be ramped up to production volumes with an acceptable yield. The goal of the productization phase is to reach mass production volumes at an acceptable cost.

Sustaining

Once a design is mature and has reached mass production it must be sustained. The process must be continually monitored and problems dealt with quickly to avoid a significant impact on production volumes. The goal of sustaining is to maintain production volumes and continually reduce costs until the product reaches end of life.

Semiconductor Device Fabrication

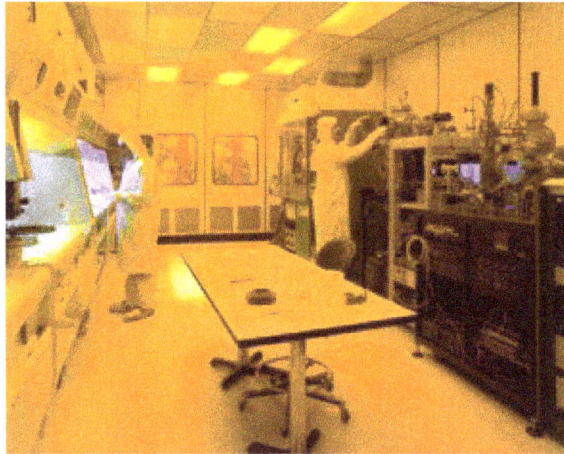

NASA's Glenn Research Center clean room.

Semiconductor device fabrication is the process used to create the integrated circuits that are present in everyday electrical and electronic devices. It is a multiple-step sequence of photo lithographic and chemical processing steps during which electronic circuits are gradually created on a wafer made of pure semiconducting material. Silicon is almost always used, but various compound semiconductors are used for specialized applications.

The entire manufacturing process, from start to packaged chips ready for shipment, takes six to eight weeks and is performed in highly specialized facilities referred to as fabs.

History

When feature widths were far greater than about 10 micrometres, purity was not the issue that it is today in device manufacturing. As devices became more integrated, cleanrooms became even cleaner. Today, the fabs are pressurized with filtered air to remove even the smallest particles, which could come to rest on the wafers and contribute to defects. The workers in a semiconductor fabrication facility are required to wear cleanroom suits to protect the devices from human contamination.

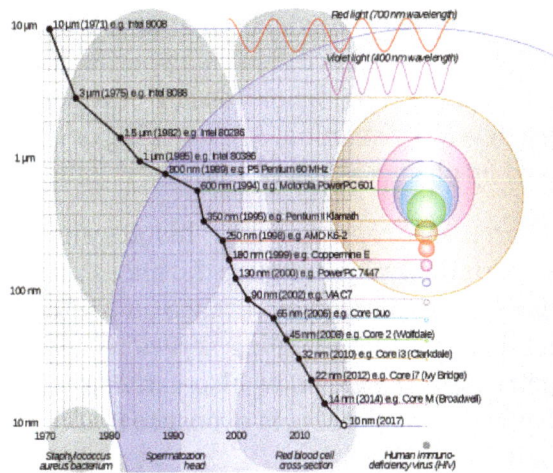

Progress of miniaturisation, and comparison of sizes of semiconductor manufacturing process nodes with some microscopic objects and visible light wavelengths.

Semiconductor device manufacturing has spread from Texas and California in the 1960s to the rest of the world, including Europe, the Middle East, and Asia. It is a global business today. The leading semiconductor manufacturers typically have facilities all over the world. Intel, the world's largest manufacturer, has facilities in Europe and Asia as well as the U.S. Other top manufacturers include Taiwan Semiconductor Manufacturing Company (Taiwan), United Microelectronics Corporation (Taiwan), STMicroelectronics (Europe), Analog Devices (US), Integrated Device Technology (US), Atmel (US/Europe), Freescale Semiconductor (US), Samsung (Korea), Texas Instruments (US), IBM (US), GlobalFoundries (Germany, Singapore, US), Toshiba (Japan), NEC Electronics (Japan), Infineon (Europe, US, Asia), Renesas (Japan), Fujitsu (Japan/US), NXP Semiconductors (Europe, Asia and US), Micron Technology (US), Hynix (Korea), and SMIC (China).

Wafers

A typical wafer is made out of extremely pure silicon that is grown into mono-crystalline cylindrical ingots (boules) up to 300 mm (slightly less than 12 inches) in diameter using the Czochralski process. These ingots are then sliced into wafers about 0.75 mm thick and polished to obtain a very regular and flat surface.

Processing

In semiconductor device fabrication, the various processing steps fall into four general categories: deposition, removal, patterning, and modification of electrical properties.

- *Deposition* is any process that grows, coats, or otherwise transfers a material onto the wafer. Available technologies include physical vapor deposition (PVD), chemical vapor deposition (CVD), electrochemical deposition (ECD), molecular beam epitaxy (MBE) and more recently, atomic layer deposition (ALD) among others.

- *Removal* is any process that removes material from the wafer; examples include etch processes (either wet or dry) and chemical-mechanical planarization (CMP).

- *Patterning* is the shaping or altering of deposited materials, and is generally referred to as lithography. For example, in conventional lithography, the wafer is coated with a chemical called a *photoresist*; then, a machine called a *stepper* focuses, aligns, and moves a mask, exposing select portions of the wafer below to short wavelength light; the exposed regions are washed away by a developer solution. After etching or other processing, the remaining photoresist is removed by plasma ashing.

- *Modification of electrical properties* has historically entailed doping transistor *sources* and *drains* (originally by diffusion furnaces and later by ion implantation). These doping processes are followed by furnace annealing or, in advanced devices, by rapid thermal annealing (RTA); annealing serves to activate the implanted dopants. Modification of electrical properties now also extends to the reduction of a material's dielectric constant in low-k insulators via exposure to ultraviolet light in UV processing (UVP).

Modern chips have up to eleven metal levels produced in over 300 sequenced processing steps.

Front-end-of-line (FEOL) Processing

FEOL processing refers to the formation of the transistors directly in the silicon. The raw wafer is engineered by the growth of an ultrapure, virtually defect-free silicon layer through epitaxy. In the most advanced logic devices, *prior* to the silicon epitaxy step, tricks are performed to improve the performance of the transistors to be built. One method involves introducing a *straining step* wherein a silicon variant such as silicon-germanium (SiGe) is deposited. Once the epitaxial silicon is deposited, the crystal lattice becomes stretched somewhat, resulting in improved electronic mobility. Another method, called *silicon on insulator* technology involves the insertion of an insulating layer between the raw silicon wafer and the thin layer of subsequent silicon epitaxy. This method results in the creation of transistors with reduced parasitic effects.

Gate Oxide and Implants

Front-end surface engineering is followed by growth of the gate dielectric (traditionally silicon dioxide), patterning of the gate, patterning of the source and drain regions, and subsequent implantation or diffusion of dopants to obtain the desired complementary electrical properties. In dynamic random-access memory (DRAM) devices, storage capacitors are also fabricated at this time, typically stacked above the access transistor (the now defunct DRAM manufacturer Qimonda implemented these capacitors with trenches etched deep into the silicon surface).

Back-end-of-line (BEOL) Processing

Metal Layers

Once the various semiconductor devices have been created, they must be interconnected to form the desired electrical circuits. This occurs in a series of wafer processing steps collectively referred to as BEOL. BEOL processing involves creating metal interconnecting wires that are isolated by dielectric layers. The insulating material has traditionally been a form of SiO_2 or a silicate glass, but recently new low dielectric constant materials are being used (such as silicon oxycarbide), typically providing dielectric constants around 2.7 (compared to 3.9 for SiO_2), although materials with constants as low as 2.2 are being offered to chipmakers.

Interconnect

Synthetic detail of a standard cell through four layers of planarized copper interconnect, down to the polysilicon (pink), wells (greyish) and substrate (green).

Historically, the metal wires have been composed of aluminum. In this approach to wiring (often called *subtractive aluminum*), blanket films of aluminum are deposited first, patterned, and then etched, leaving isolated wires. Dielectric material is then deposited over the exposed wires. The various metal layers are interconnected by etching holes (called *"vias"*) in the insulating material and then depositing tungsten in them with a CVD technique; this approach is still used in the fabrication of many memory chips such as dynamic random-access memory (DRAM), because the number of interconnect levels is small (currently no more than four).

More recently, as the number of interconnect levels for logic has substantially increased due to the large number of transistors that are now interconnected in a modern microprocessor, the timing delay in the wiring has become so significant as to prompt a change in wiring material (from aluminum to copper layer) and a change in dielectric material (from silicon dioxides to newer low-K insulators). This performance enhancement also comes at a reduced cost via damascene processing, which eliminates processing steps. As the number of interconnect levels increases, planarization of the previous layers is required to ensure a flat surface prior to subsequent lithography. Without it, the levels would become increasingly crooked, extending outside the depth of focus of available lithography, and thus interfering with the ability to pattern. CMP (chemical-mechanical planarization) is the primary processing method to achieve such planarization, although dry *etch back* is still sometimes employed when the number of interconnect levels is no more than three.

Wafer Test

The highly serialized nature of wafer processing has increased the demand for metrology in between the various processing steps. For example, thin film metrology based on ellipsometry or reflectometry is used to tightly control the thickness of gate oxide, as well as the thickness, refractive index and extinction coefficient of photoresist and other coatings. Wafer test metrology equipment is used to verify that the wafers haven't been damaged by previous processing steps up until testing; if too many dies on one wafer have failed, the entire wafer is scrapped to avoid the costs of further processing. Virtual metrology has been used to predict wafer properties based on statistical methods without performing the physical measurement itself.

Device Test

Once the front-end process has been completed, the semiconductor devices are subjected to a variety of electrical tests to determine if they function properly. The proportion of devices on the wafer found to perform properly is referred to as the yield. Manufacturers are typically secretive about their yields, but it can be as low as 30%. Process variation is one among many reasons for low yield.

The fab tests the chips on the wafer with an electronic tester that presses tiny probes against the chip. The machine marks each bad chip with a drop of dye. Currently, electronic dye marking is possible if wafer test data is logged into a central computer database and chips are "binned" (i.e. sorted into virtual bins) according to the predetermined test limits. The resulting binning data can be graphed, or logged, on a wafer map to trace manufacturing defects and mark bad chips. This map can also be used during wafer assembly and packaging.

Chips are also tested again after packaging, as the bond wires may be missing, or analog performance may be altered by the package. This is referred to as the "final test".

Usually, the fab charges for testing time, with prices in the order of cents per second. Testing times vary from a few milliseconds to a couple of seconds, and the test software is optimized for reduced testing time. Multiple chip (multi-site) testing is also possible, because many testers have the resources to perform most or all of the tests in parallel.

Chips are often designed with "testability features" such as scan chains or a "built-in self-test" to speed testing, and reduce testing costs. In certain designs that use specialized analog fab processes, wafers are also laser-trimmed during the testing, in order to achieve tightly-distributed resistance values as specified by the design.

Good designs try to test and statistically manage *corners* (extremes of silicon behavior caused by a high operating temperature combined with the extremes of fab processing steps). Most designs cope with at least 64 corners.

Die Preparation

Once tested, a wafer is typically reduced in thickness before the wafer is scored and then broken into individual dice, a process known as wafer dicing. Only the good, unmarked chips are packaged.

Packaging

Plastic or ceramic packaging involves mounting the die, connecting the die pads to the pins on the package, and sealing the die. Tiny wires are used to connect the pads to the pins. In the old days, wires were attached by hand, but now specialized machines perform the task. Traditionally, these wires have been composed of gold, leading to a lead frame (pronounced "leed frame") of solder-plated copper; lead is poisonous, so lead-free "lead frames" are now mandated by RoHS.

Chip scale package (CSP) is another packaging technology. A plastic dual in-line package, like most packages, is many times larger than the actual die hidden inside, whereas CSP chips are nearly the size of the die; a CSP can be constructed for each die *before* the wafer is diced.

The packaged chips are retested to ensure that they were not damaged during packaging and that the die-to-pin interconnect operation was performed correctly. A laser then etches the chip's name and numbers on the package.

List of Steps

This is a list of processing techniques that are employed numerous times throughout the construction of a modern electronic device; this list does not necessarily imply a specific order.

- Wafer processing
  - Wet cleans
    - ☐ Cleaning by solvents such as acetone, trichloroethylene
    - ☐ Piranha solution
    - ☐ RCA clean
  - Photolithography
  - Ion implantation (in which dopants are embedded in the wafer creating regions of increased (or decreased) conductivity)
  - Dry etching
  - Wet etching
  - Plasma ashing
  - Thermal treatments
    - ☐ Rapid thermal anneal
    - ☐ Furnace anneals
    - ☐ Thermal oxidation
  - Chemical vapor deposition (CVD)
  - Physical vapor deposition (PVD)
  - Molecular beam epitaxy (MBE)
  - Electrochemical deposition (ECD).
  - Chemical-mechanical planarization (CMP)
  - Wafer testing (where the electrical performance is verified)

- o Wafer backgrinding (to reduce the thickness of the wafer so the resulting chip can be put into a thin device like a smartcard or PCMCIA card.)
- Die preparation
  - o Wafer mounting
  - o Die cutting
- IC packaging
  - o Die attachment
  - o IC bonding
    - ☐ Wire bonding
    - ☐ Thermosonic bonding
    - ☐ Flip chip
    - ☐ Wafer bonding
    - ☐ Tape Automated Bonding (TAB)
  - o IC encapsulation
    - ☐ Baking
    - ☐ Plating
    - ☐ Lasermarking
    - ☐ Trim and form
- IC testing

Hazardous Materials

Many toxic materials are used in the fabrication process. These include:

- poisonous elemental dopants, such as arsenic, antimony, and phosphorus.
- poisonous compounds, such as arsine, phosphine, and silane.
- highly reactive liquids, such as hydrogen peroxide, fuming nitric acid, sulfuric acid, and hydrofluoric acid.

It is vital that workers should not be directly exposed to these dangerous substances. The high degree of automation common in the IC fabrication industry helps to reduce the risks of exposure. Most fabrication facilities employ exhaust management systems, such as wet scrubbers, combustors, heated absorber cartridges, etc., to control the risk to workers and to the environment.

Wafer-scale Integration

Wafer-scale integration, WSI for short, is a rarely used system of building very-large integrated circuit networks that use an entire silicon wafer to produce a single "super-chip". Through a combination of large size and reduced packaging, WSI could lead to dramatically reduced costs for some systems, notably massively parallel supercomputers. The name is taken from the term very-large-scale integration, the current state of the art when WSI was being developed.

The Concept

To understand WSI, one has to consider the normal chip-making process. A single large cylindrical crystal of silicon is produced and then cut into disks known as wafers. The wafers are then cleaned and polished in preparation for the fabrication process. A photographic process is used to pattern the surface where material ought to be deposited on top of the wafer and where not to. The desired material is deposited and the photographic mask is removed for the next layer. From then on the wafer is repeatedly processed in this fashion, putting on layer after layer of circuitry on the surface.

Multiple copies of these patterns are deposited on the wafer in a grid fashion across the surface of the wafer. After all the possible locations are patterned, the wafer surface appears like a sheet of graph paper, with grid lines delineating the individual chips. Each of these grid locations is tested for manufacturing defects by automated equipment. Those locations that are found to be defective are recorded and marked with a dot of paint (this process is referred to as "inking a die" however modern wafer fabrication no longer requires physical markings to identify defective die). The wafer is then sawed apart to cut out the individual chips. Those defective chips are thrown away, or recycled, while the working chips are placed into packaging and re-tested for any damage that might occur during the packaging process.

Flaws on the surface of the wafers and problems during the layering/depositing process are impossible to avoid, and cause some of the individual chips to be defective. The revenue from the remaining working chips has to pay for the entire cost of the wafer and its processing, including those discarded defective chips. Thus, the higher number of working chips or higher *yield*, the lower the cost of each individual chip. In order to maximize yield one wants to make the chips as small as possible, so that a higher number of working chips can be obtained per wafer.

The vast majority of the cost of fabrication (typically 30%-50%) is related to testing and packaging the individual chips. Further cost is associated with connecting the chips into an integrated system (usually via a printed circuit board). Wafer-scale integration seeks to reduce this cost, as well as improve performance, by building larger chips in a single package – in principle, chips as large as a full wafer.

Of course this is not easy, since given the flaws on the wafers a single large design printed onto a wafer would almost always not work. It has been an ongoing goal to develop methods to handle faulty areas of the wafers through logic, as opposed to sawing them out of the wafer. Generally, this approach uses a grid pattern of sub-circuits and "rewires" around the damaged areas using appropriate logic. If the resulting wafer has enough working sub-circuits, it can be used despite faults.

Production Attempts

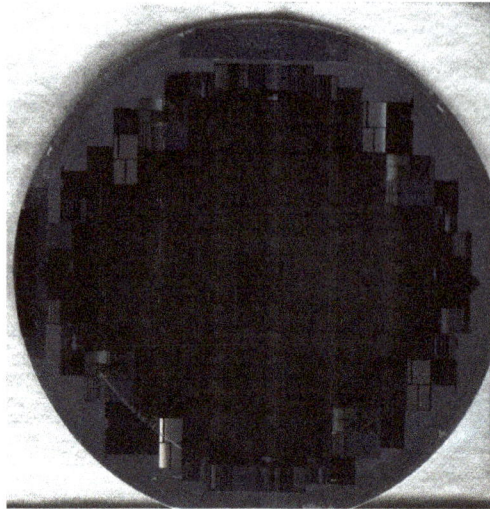

Early WSI attempt by Trilogy Systems.

Many companies attempted to develop WSI production systems in the 1970s and 80s, but all failed. TI and ITT both saw it as a way to develop complex pipelined microprocessors and re-enter a market where they were losing ground, but neither released any products.

Gene Amdahl also attempted to develop WSI as a method of making a supercomputer, starting Trilogy Systems in 1980 and garnering investments from Groupe Bull, Sperry Rand and Digital Equipment Corporation, who (along with others) provided an estimated $230 million in financing. The design called for a 2.5" square chip with 1200 pins on the bottom.

The effort was plagued by a series of disasters, including floods which delayed the construction of the plant and later ruined the clean-room interior. After burning through about 1/3 of the capital with nothing to show for it, Amdahl eventually declared the idea would only work with a 99.99% yield, which wouldn't happen for 100 years. He used Trilogy's remaining seed capital to buy Elxsi, a maker of VAX-compatible machines, in 1985. The Trilogy efforts were eventually ended and "became" Elxsi.

System on a Chip

The Raspberry Pi uses a system on a chip as a fully-contained micro computer.

A system on a chip or system on chip (SoC or SOC) is an integrated circuit (IC) that integrates all components of a computer or other electronic system into a single chip. It may contain digital, analog, mixed-signal, and often radio-frequency functions—all on a single chip substrate. SoCs are very common in the mobile electronics market because of their low power-consumption. A typical application is in the area of embedded systems.

The contrast with a microcontroller is one of degree. Microcontrollers typically have under 100 kB of RAM (often just a few kilobytes) and often really *are* single-chip-systems, whereas the term SoC is typically used for more powerful processors, such as on smartphones. SoCs are capable of running software such as the desktop versions of Windows and Linux, which need external memory chips (flash, RAM) to be useful, and which are used with various external peripherals. In short, for larger systems, the term *system on a chip* is hyperbole, indicating technical direction more than reality: a high degree of chip integration, leading toward reduced manufacturing costs, and the production of smaller systems. Many systems are too complex to fit on just one chip built with a processor optimized for just one of the system's tasks.

When it is not feasible to construct a SoC for a particular application, an alternative is a system in package (SiP) comprising a number of chips in a single package. In large volumes, SoC is believed to be more cost-effective than SiP since it increases the yield of the fabrication and because its packaging is simpler.

Another option, as seen for example in higher-end cell phones, is package on package stacking during board assembly. The SoC includes processors and numerous digital peripherals, and comes in a ball grid package with lower and upper connections. The lower balls connect to the board and various peripherals, with the upper balls in a ring holding the memory buses used to access NAND flash and DDR2 RAM. Memory packages could come from multiple vendors.

AMD Am286ZX/LX, SoC based on 80286

Structure

A typical SoC consists of:

- a microcontroller, microprocessor or digital signal processor (DSP) core – multiprocessor SoCs (MPSoC) having more than one processor core

- memory blocks including a selection of ROM, RAM, EEPROM and flash memory

- timing sources including oscillators and phase-locked loops

- peripherals including counter-timers, real-time timers and power-on reset generators

- external interfaces, including industry standards such as USB, FireWire, Ethernet, USART, SPI

- analog interfaces including ADCs and DACs

- voltage regulators and power management circuits

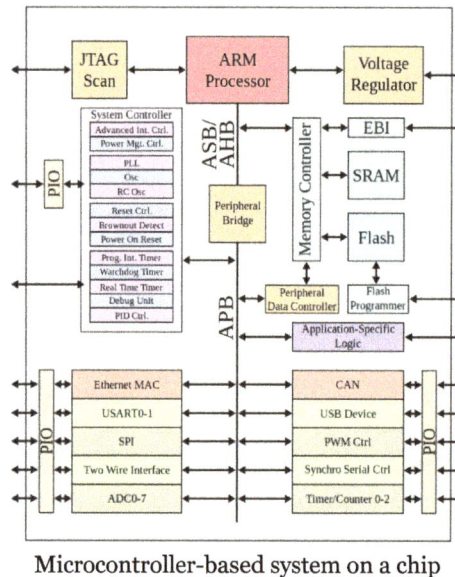

Microcontroller-based system on a chip

A bus – either proprietary or industry-standard such as the AMBA bus from ARM Holdings – connects these blocks. DMA controllers route data directly between external interfaces and memory, bypassing the processor core and thereby increasing the data throughput of the SoC.

Design Flow

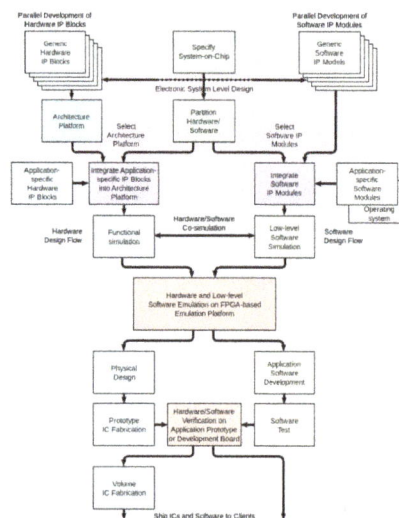

System-on-a-chip design flow

A SoC consists of both the hardware, described above, and the software controlling the microcontroller, microprocessor or DSP cores, peripherals and interfaces. The design flow for a SoC aims to develop this hardware and software in parallel.

Most SoCs are developed from pre-qualified hardware blocks for the hardware elements described above, together with the software drivers that control their operation. Of particular importance are the protocol stacks that drive industry-standard interfaces like USB. The hardware blocks are put together using CAD tools; the software modules are integrated using a software-development environment.

Once the architecture of the SoC has been defined, any new hardware elements are written in an abstract language termed RTL which defines the circuit behaviour. These elements are connected together in the same RTL language to create the full SoC design.

Chips are verified for logical correctness before being sent to foundry. This process is called functional verification and it accounts for a significant portion of the time and energy expended in the chip design life cycle (although the often quoted figure of 70% is probably an exaggeration). With the growing complexity of chips, hardware verification languages like SystemVerilog, SystemC, e, and OpenVera are being used. Bugs found in the verification stage are reported to the designer.

Traditionally, engineers have employed simulation acceleration, emulation and/or an FPGA prototype to verify and debug both hardware and software for SoC designs prior to tapeout. With high capacity and fast compilation time, acceleration and emulation are powerful technologies that provide wide visibility into systems. Both technologies, however, operate slowly, on the order of MHz, which may be significantly slower – up to 100 times slower – than the SoC's operating frequency. Acceleration and emulation boxes are also very large and expensive at over US$1,000,000. FPGA prototypes, in contrast, use FPGAs directly to enable engineers to validate and test at, or close to, a system's full operating frequency with real-world stimuli. Tools such as Certus are used to insert probes in the FPGA RTL that make signals available for observation. This is used to debug hardware, firmware and software interactions across multiple FPGAs with capabilities similar to a logic analyzer.

In parallel, the hardware elements are grouped and passed through a process of logic synthesis, during which performance constraints, such as operational frequency and expected signal delays, are applied. This generates a logical netlist which is a file describing the circuit as a collection of connected silicon gate elements from a library provided by the silicon manufacturer.

This netlist is used as the basis for the physical design (place and route) flow to convert the designers' intent into the polygonal design of the SoC. Throughout this conversion process, the design is analysed with static timing modelling, simulation and other tools to ensure that it meets the specified operational parameters such as frequency, power consumption and dissipation, functional integrity vs. the RTL and electrical integrity.

When all known bugs have been rectified and these have been re-verified and all physical design checks are done, the physical design files describing each layer of the chip are sent to the foundry's mask shop where a full set of glass lithographic masks will be etched. These are sent to the wafer fabrication plant to create the SoC dice before packaging and testing.

Fabrication

SoCs can be fabricated by several technologies, including:

- Full custom

- Standard cell

- Field-programmable gate array (FPGA)

SoC designs usually consume less power and have a lower cost and higher reliability than the multi-chip systems that they replace. And with fewer packages in the system, assembly costs are reduced as well.

However, like most VLSI designs, the total cost is higher for one large chip than for the same functionality distributed over several smaller chips, because of lower yields and higher non-recurring engineering costs.

Benchmarks

SoC research and development often compares many options. Benchmarks, such as COSMIC, are developed to help such evaluations.

Three-dimensional Integrated Circuit

In microelectronics, a three-dimensional integrated circuit (3D IC) is an integrated circuit manufactured by stacking silicon wafers and/or dies and interconnecting them vertically using through-silicon vias (TSVs) so that they behave as a single device to achieve performance improvements at reduced power and smaller footprint than conventional two dimensional processes. 3D IC is just one of a host of 3D integration schemes that exploit the z-direction to achieve electrical performance benefits. They can be classified by their level of interconnect hierarchy at the global (package), intermediate (bond pad) and local (transistor) level In general, 3D integration is a broad term that includes such technologies as 3D wafer-level packaging (3DWLP); 2.5D and 3D interposer-based integration; 3D stacked ICs (3D-SICs), monolithic 3D ICs; 3D heterogeneous integration; and 3D systems integration. International organizations such as the Jisso Technology Roadmap Committee (JIC) and the International Technology Roadmap for Semiconductors (ITRS) have worked to classify the various 3D integration technologies to further the establishment of standards and roadmaps of 3D integration.

3D ICs vs. 3D Packaging

3D Packaging refers to 3D integration schemes that rely on traditional methods of interconnect such as wire bonding and flip chip to achieve vertical stacks. 3D packaging can be disseminated further into 3D system in package (3D SiP) and 3D wafer level package (3D WLP). Stacked memory die interconnected with wire bonds, and package on package (PoP) configurations interconnected with either wire bonds, or flip chips are 3D SiPs that have been in mainstream manufacturing for some time and have a well established infrastructure. PoP is used for vertically integrating disparate technologies such as 3D WLP uses wafer level processes such as redistribution layers (RDL) and wafer bumping processes to form interconnects.

2.5D interposer is also a 3D WLP that interconnects die side-side on a silicon, glass or organic interposer using TSVs and RDL. In all types of 3D Packaging, chips in the package communicate using off-chip signaling, much as if they were mounted in separate packages on a normal circuit board.

3D ICs can be divided into 3D Stacked ICs (3D SIC), which refers to stacking IC chips using TSV interconnects, and monolithic 3D ICs, which use fab processes to realize 3D interconnects at the local levels of the on-chip wiring hierarchy as set forth by the ITRS, this results in direct vertical interconnects between device layers. The first examples of a monolithic approach are seen in Samsung's 3D VNAND devices.

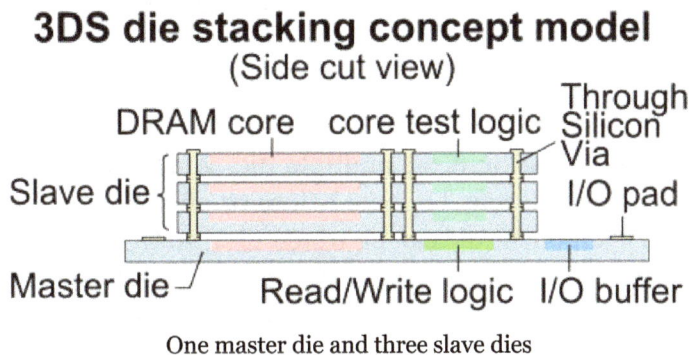

One master die and three slave dies

3D SiCs

The digital electronics market requires a higher density semiconductor memory chip to cater to recently released CPU components, and the multiple die stacking technique has been suggested as a solution to this problem. JEDEC disclosed the upcoming DRAM technology includes the "3D SiC" die stacking plan at "Server Memory Forum", November 1–2, 2011, Santa Clara, CA. In August 2014, Samsung started producing 64GB DRAM modules for servers based on emerging DDR4 (double-data rate 4) memory using 3D TSV package technology. Newer proposed standards for 3D stacked DRAM include Wide I/O, Wide I/O 2, Hybrid Memory Cube, High Bandwidth Memory.

Monolithic 3D ICs

Monolithic 3D ICs are built in layers on a single semiconductor wafer, which is then diced into 3D ICs. There is only one substrate, hence no need for aligning, thinning, bonding, or through-silicon vias. Process temperature limitations are addressed by partitioning the transistor fabrication to two phases. A high temperature phase which is done before layer transfer follow by a layer transfer use ion-cut, also known as layer transfer, which has been used to produce Silicon on Insulator (SOI) wafers for the past two decades. Multiple thin (10s–100s nanometer scale) layers of virtually defect-free Silicon can be created by utilizing low temperature (<400℃) bond and cleave techniques, and placed on top of active transistor circuitry. Follow by finalizing the transistors using etch and deposition processes. This monolithic 3D IC technology has been researched at Stanford University under a DARPA-sponsored grant.

CEA-Leti is also developing monolithic 3D IC approaches, called sequential 3D IC. In 2014, the French research institute introduced its CoolCube™, a low-temperature process flow that provides a true path to 3DVLSI. At Stanford University, researchers are designing monolithic 3D ICs using carbon nanotube (CNT) structures vs. silicon using a wafer-scale low temperature CNT transfer processes that can be done at 120℃.

In general, monolithic 3D ICs are still a developing technology and are considered by most to be several years away from production.

Manufacturing Technologies for 3D SiCs

As of 2014, a number of memory products such as High Bandwidth Memory (HBM) and the Hybrid Memory Cube have been launched that implement 3D IC stacking with TSVs. There are a number of key stacking approaches being implemented and explored. These include die-to-die, die-to-wafer, and wafer-to-wafer.

Die-to-Die

> Electronic components are built on multiple die, which are then aligned and bonded. Thinning and TSV creation may be done before or after bonding. One advantage of die-to-die is that each component die can be tested first, so that one bad die does not ruin an entire stack. Moreover, each die in the 3D IC can be binned beforehand, so that they can be mixed and matched to optimize power consumption and performance (e.g. matching multiple dice from the low power process corner for a mobile application).

Die-to-Wafer

> Electronic components are built on two semiconductor wafers. One wafer is diced; the singulated dice are aligned and bonded onto die sites of the second wafer. As in the wafer-on-wafer method, thinning and TSV creation are performed either before or after bonding. Additional die may be added to the stacks before dicing.

Wafer-to-Wafer

> Electronic components are built on two or more semiconductor wafers, which are then aligned, bonded, and diced into 3D ICs. Each wafer may be thinned before or after bonding. Vertical connections are either built into the wafers before bonding or else created in the stack after bonding. These "through-silicon vias" (TSVs) pass through the silicon substrate(s) between active layers and/or between an active layer and an external bond pad. Wafer-to-wafer bonding can reduce yields, since if any 1 of N chips in a 3D IC are defective, the entire 3D IC will be defective. Moreover, the wafers must be the same size, but many exotic materials (e.g. III-Vs) are manufactured on much smaller wafers than CMOS logic or DRAM (typically 300 mm), complicating heterogeneous integration.

Benefits of 3D ICs

While traditional CMOS scaling processes improves signal propagation speed, scaling from current manufacturing and chip-design technologies is becoming more difficult and costly, in part because of power-density constraints, and in part because interconnects do not become faster while transistors do. 3D ICs address the scaling challenge by stacking 2D dies and connecting them in the 3rd dimension. This promises to speed up communication between layered chips, compared to planar layout. 3D ICs promise many significant benefits, including:

Footprint

> More functionality fits into a small space. This extends Moore's law and enables a new generation of tiny but powerful devices.

Cost

> Partitioning a large chip into multiple smaller dies with 3D stacking can improve the yield and reduce the fabrication cost if individual dies are tested separately.

Heterogeneous integration

> Circuit layers can be built with different processes, or even on different types of wafers. This means that components can be optimized to a much greater degree than if they were built together on a single wafer. Moreover, components with incompatible manufacturing could be combined in a single 3D IC.

Shorter interconnect

> The average wire length is reduced. Common figures reported by researchers are on the order of 10–15%, but this reduction mostly applies to longer interconnect, which may affect circuit delay by a greater amount. Given that 3D wires have much higher capacitance than conventional in-die wires, circuit delay may or may not improve.

Power

> Keeping a signal on-chip can reduce its power consumption by 10–100 times. Shorter wires also reduce power consumption by producing less parasitic capacitance. Reducing the power budget leads to less heat generation, extended battery life, and lower cost of operation.

Design

> The vertical dimension adds a higher order of connectivity and offers new design possibilities.

Circuit security

> Security through obscurity. The stacked structure complicates attempts to reverse engineer the circuitry. Sensitive circuits may also be divided among the layers in such a way as to obscure the function of each layer.

Bandwidth

> 3D integration allows large numbers of vertical vias between the layers. This allows construction of wide bandwidth buses between functional blocks in different layers. A typical example would be a processor+memory 3D stack, with the cache memory stacked on top of

the processor. This arrangement allows a bus much wider than the typical 128 or 256 bits between the cache and processor. Wide buses in turn alleviate the memory wall problem.

Challenges

Because this technology is new it carries new challenges, including:

Cost

> While cost is a benefit when compared with scaling, it has also been identified as a challenge to the commercialization of 3D ICs in mainstream consumer applications. However, work is being done to address this. Although 3D technology is new and fairly complex, the cost of the manufacturing process is surprisingly straightforward when broken down into the activities that build up the entire process. By analyzing the combination of activities that lay at the base, cost drivers can be identified. Once the cost drivers are identified, it becomes a less complicated endeavor to determine where the majority of cost comes from and, more importantly, where cost has the potential to be reduced.

Yield

> Each extra manufacturing step adds a risk for defects. In order for 3D ICs to be commercially viable, defects could be repaired or tolerated, or defect density can be improved.

Heat

> Heat building up within the stack must be dissipated. This is an inevitable issue as electrical proximity correlates with thermal proximity. Specific thermal hotspots must be more carefully managed.

Design complexity

> Taking full advantage of 3D integration requires sophisticated design techniques and new CAD tools.

TSV-introduced overhead

> TSVs are large compared to gates and impact floorplans. At the 45 nm technology node, the area footprint of a 10μm x 10μm TSV is comparable to that of about 50 gates. Furthermore, manufacturability demands landing pads and keep-out zones which further increase TSV area footprint. Depending on the technology choices, TSVs block some subset of layout resources. Via-first TSVs are manufactured before metallization, thus occupy the device layer and result in placement obstacles. Via-last TSVs are manufactured after metallization and pass through the chip. Thus, they occupy both the device and metal layers, resulting in placement and routing obstacles. While the usage of TSVs is generally expected to reduce wirelength, this depends on the number of TSVs and their characteristics. Also, the granularity of inter-die partitioning impacts wirelength. It typically decreases for moderate (blocks with 20-100 modules) and coarse (block-level partitioning) granularities, but increases for fine (gate-level partitioning) granularities.

Testing

> To achieve high overall yield and reduce costs, separate testing of independent dies is essential. However, tight integration between adjacent active layers in 3D ICs entails a significant amount of interconnect between different sections of the same circuit module that were partitioned to different dies. Aside from the massive overhead introduced by required TSVs, sections of such a module, e.g., a multiplier, cannot be independently tested by conventional techniques. This particularly applies to timing-critical paths laid out in 3D.

Lack of standards

> There are few standards for TSV-based 3D IC design, manufacturing, and packaging, although this issue is being addressed. In addition, there are many integration options being explored such as via-last, via-first, via-middle; interposers or direct bonding; etc.

Heterogeneous integration supply chain

> In heterogeneously integrated systems, the delay of one part from one of the different parts suppliers delays the delivery of the whole product, and so delays the revenue for each of the 3D IC part suppliers.

Lack of clearly defined ownership

> It is unclear who should own the 3D IC integration and packaging/assembly. It could be assembly houses like ASE or the product OEMs.

Design Styles

Depending on partitioning granularity, different design styles can be distinguished. Gate-level integration faces multiple challenges and currently appears less practical than block-level integration.

Gate-level integration

> This style partitions standard cells between multiple dies. It promises wirelength reduction and great flexibility. However, wirelength reduction may be undermined unless modules of certain minimal size are preserved. On the other hand, its adverse effects include the massive number of necessary TSVs for interconnects. This design style requires 3D place-and-route tools, which are unavailable yet. Also, partitioning a design block across multiple dies implies that it cannot be fully tested before die stacking. After die stacking (post-bond testing), a single failed die can render several good dies unusable, undermining yield. This style also amplifies the impact of process variation, especially inter-die variation. In fact, a 3D layout may yield more poorly than the same circuit laid out in 2D, contrary to the original promise of 3D IC integration. Furthermore, this design style requires to redesign available Intellectual Property, since existing IP blocks and EDA tools do not provision for 3D integration.

Block-level integration

> This style assigns entire design blocks to separate dies. Design blocks subsume most of the netlist connectivity and are linked by a small number of global interconnects. Therefore,

block-level integration promises to reduce TSV overhead. Sophisticated 3D systems combining heterogeneous dies require distinct manufacturing processes at different technology nodes for fast and low-power random logic, several memory types, analog and RF circuits, etc. Block-level integration, which allows separate and optimized manufacturing processes, thus appears crucial for 3D integration. Furthermore, this style might facilitate the transition from current 2D design towards 3D IC design. Basically, 3D-aware tools are only needed for partitioning and thermal analysis. Separate dies will be designed using (adapted) 2D tools and 2D blocks. This is motivated by the broad availability of reliable IP blocks. It is more convenient to use available 2D IP blocks and to place the mandatory TSVs in the unoccupied space between blocks instead of redesigning IP blocks and embedding TSVs. Design-for-testability structures are a key component of IP blocks and can therefore be used to facilitate testing for 3D ICs. Also, critical paths can be mostly embedded within 2D blocks, which limits the impact of TSV and inter-die variation on manufacturing yield. Finally, modern chip design often requires last-minute engineering changes. Restricting the impact of such changes to single dies is essential to limit cost.

Notable 3D Chips

In 2004 Tezzaron Semiconductor built working 3D devices from six different designs. The chips were built in two layers with "via-first" tungsten TSVs for vertical interconnection. Two wafers were stacked face-to-face and bonded with a copper process. The top wafer was thinned and the two-wafer stack was then diced into chips. The first chip tested was a simple memory register, but the most notable of the set was an 8051 processor/memory stack that exhibited much higher speed and lower power consumption than an analogous 2D assembly.

In 2004, Intel presented a 3D version of the Pentium 4 CPU. The chip was manufactured with two dies using face-to-face stacking, which allowed a dense via structure. Backside TSVs are used for I/O and power supply. For the 3D floorplan, designers manually arranged functional blocks in each die aiming for power reduction and performance improvement. Splitting large and high-power blocks and careful rearrangement allowed to limit thermal hotspots. The 3D design provides 15% performance improvement (due to eliminated pipeline stages) and 15% power saving (due to eliminated repeaters and reduced wiring) compared to the 2D Pentium 4.

The Teraflops Research Chip introduced in 2007 by Intel is an experimental 80-core design with stacked memory. Due to the high demand for memory bandwidth, a traditional I/O approach would consume 10 to 25 W. To improve upon that, Intel designers implemented a TSV-based memory bus. Each core is connected to one memory tile in the SRAM die with a link that provides 12 GB/s bandwidth, resulting in a total bandwidth of 1 TB/s while consuming only 2.2 W.

An academic implementation of a 3D processor was presented in 2008 at the University of Rochester by Professor Eby Friedman and his students. The chip runs at a 1.4 GHz and it was designed for optimized vertical processing between the stacked chips which gives the 3D processor abilities that the traditional one layered chip could not reach. One challenge in manufacturing of the three-dimensional chip was to make all of the layers work in harmony without any obstacles that would interfere with a piece of information traveling from one layer to another.

In ISSCC 2012, two 3D-IC-based multi-core designs using GlobalFoundries' 130 nm process and

Tezzaron's FaStack technology were presented and demonstrated. 3D-MAPS, a 64 custom core implementation with two-logic-die stack was demonstrated by researchers from the School of Electrical and Computer Engineering at Georgia Institute of Technology. The second prototype was from the Department of Electrical Engineering and Computer Science at University of Michigan called Centip3De, a near-threshold design based on ARM Cortex-M3 cores.

Integrated Circuit Packaging

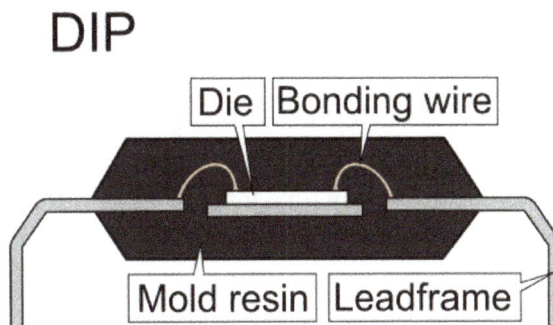

DIP

Die | Bonding wire

Mold resin | Leadframe

Cross section of a dual in-line package. This type of package houses a small semiconducting die, with nanowires attaching the die to the lead frames, allowing for electrical connections to be made to a PCB.

In electronics manufacturing, integrated circuit packaging is the final stage of semiconductor device fabrication, in which the tiny block of semiconducting material is encapsulated in a supporting case that prevents physical damage and corrosion. The case, known as a "package", supports the electrical contacts which connect the device to a circuit board.

In the integrated circuit industry, the process is often referred to as packaging. Other names include semiconductor device assembly, assembly, encapsulation or sealing.

The packaging stage is followed by testing of the integrated circuit.

The term is sometimes confused with electronic packaging, which is the mounting and interconnecting of integrated circuits (and other components) onto printed-circuit boards.

Design Considerations

Electrical

The current-carrying traces that run out of the die, through the package, and into the printed circuit board (PCB) have very different electrical properties compared to on-chip signals. They require special design techniques and need much more electric power than signals confined to the chip itself. Therefore, it is important that the materials used as electrical contacts exhibit characteristics like low resistance, low capacitance and low inductance. Both the structure and materials must prioritize signal transmission properties, while minimizing any parasitic elements that could negatively affect the signal.

Controlling these characteristics is becoming increasingly important as the rest of technology begins to speed up. Packaging delays have the potential to make up almost half of a high-performance computer's delay, and this bottleneck on speed is expected to increase.

Mechanical and Thermal

The integrated circuit package is responsible for keeping the chip safe from all sorts of potential damage. The package must resist physical breakage, provide an airtight seal to keep out moisture, and also provide effective heat dissipation away from the chip. At the same time, it must have effective means of connecting to a PCB, which can change drastically depending on the package type. The materials used for the body of the package are typically either plastic or ceramic. They both can offer a high thermal conductivity and decent mechanical strength. Ceramic generally has more preferable characteristics, but is more expensive.

Increasing the surface area of the package allows for better heat transfer via convection, and some packages utilize metallic fins to enhance heat transfer even further at the cost of valuable space. Larger sizes also allow for a greater number of mechanical connections. However, these factors are balanced out by the fact that the package generally needs to be kept as small as possible.

Economic

Cost is a major limiting factor for many designs. Choices such as package material and level of precision must be balanced by the economic viability of the end product. Depending on the needs of the system, opting for lower-cost materials is often an acceptable solution to economic constraints. Typically, an inexpensive plastic package can dissipate heat up to 2W, which is sufficient for many simple applications, though a similar ceramic package can dissipate up to 50W in the same scenario. As the chips inside THE package get smaller and faster, they also tend to get hotter. As the subsequent need for more effective heat dissipation increases, the cost of packaging rises along with it. Generally, the smaller and more complex the package needs to be, the more expensive it is to manufacture.

History

Small-outline integrated circuit. This package has 16 "gull wing" leads protruding from the two long sides and a lead spacing of 0.050 inches.

The earliest integrated circuits were packaged in ceramic flat packs, which the military used for many years for their reliability and small size. Commercial circuit packaging quickly moved to the dual in-line package (DIP), first in ceramic and later in plastic. In the 1980s VLSI pin counts exceeded the practical limit for DIP packaging, leading to pin grid array (PGA) and leadless chip carrier (LCC) packages. Surface mount packaging appeared in the early 1980s and became popular in the late 1980s, using finer lead pitch with leads formed as either gull-wing or J-lead, as exemplified by small-outline integrated circuit — a carrier which occupies an area about 30 – 50% less than an equivalent DIP, with a typical thickness that is 70% less.

The next big innovation was the *area array package*, which places the interconnection terminals throughout the surface area of the package, providing a greater number of connections than previous package types where only the outer perimeter is used. The first area array package was a ceramic pin grid array package. Not long after, the plastic ball grid array (BGA), another type of area array package, became one of the most commonly used packaging techniques.

In the late 1990s, plastic quad flat pack (PQFP) and thin small-outline packages (TSOP) replaced PGA packages as the most common for high pin count devices, though PGA packages are still often used for microprocessors. However, industry leaders Intel and AMD transitioned in the 2000s from PGA packages to land grid array (LGA) packages.

Ball grid array (BGA) packages have existed since the 1970s, but evolved into Flip-chip ball grid array packages (FCBGA) in the 1990s. FCBGA packages allow for much higher pin count than any existing package types. In an FCBGA package, the die is mounted upside-down (flipped) and connects to the package balls via a substrate that is similar to a printed-circuit board rather than by wires. FCBGA packages allow an array of input-output signals (called Area-I/O) to be distributed over the entire die rather than being confined to the die periphery.

Traces out of the die, through the package, and into the printed circuit board have very different electrical properties, compared to on-chip signals. They require special design techniques and need much more electric power than signals confined to the chip itself.

Recent developments consist of stacking multiple dies in single package called SiP, for *System In Package*, or three-dimensional integrated circuit. Combining multiple dies on a small substrate, often ceramic, is called an MCM, or Multi-Chip Module. The boundary between a big MCM and a small printed circuit board is sometimes blurry.

Common Package Types

- Through-hole package
- Surface mount
- Chip carrier
- Pin grid array
- Flat package
- Small outline package
- Chip-scale package
- Ball grid array
- Transistor, diode, small pin count IC packages
- Multi-chip packages

Operations

Die attachment is the step during which a die is mounted and fixed to the package or support structure (header). For high-powered applications, the die is usually eutectic bonded onto the package, using e.g. gold-tin or gold-silicon solder (for good heat conduction). For low-cost, low-powered applications, the die is often glued directly onto a substrate (such as a printed wiring board) using an epoxy adhesive.

The following operations are performed at the packaging stage, as broken down into bonding, encapsulation, and wafer bonding steps. Note that this list is not all-inclusive and not all of these operations are performed for every package, as the process is highly dependent on the package type.

- IC Bonding
  - o Wire bonding
  - o Thermosonic Bonding
  - o Down bonding
  - o Tape-automated bonding
  - o Flip chip
  - o Quilt packaging
  - o Tab bonding
  - o Film attaching
  - o Spacer attaching
- IC encapsulation
  - o Baking
  - o Plating
  - o Lasermarking
  - o Trim and form
- Wafer bonding

Mixed-signal Integrated Circuit

A mixed-signal integrated circuit is any integrated circuit that has both analog circuits and digital circuits on a single semiconductor die.   In real-life applications mixed-signal designs are everywhere, for example, a smart mobile phone. However, it is more accurate to call them mixed-signal systems. Mixed-signal ICs also process both analog and digital signals together. For example, an analog-to-digital converter is a mixed-signal circuit. Mixed-signal circuits or systems are typically cost-effective solutions for building any modern consumer electronics applications.

Introduction

A mixed-signal system-on-a-chip (AMS-SoC) can be a combination of analog circuits, digital circuits, intrinsic mixed-signal circuits (like ADC), and embedded software.

Integrated circuits (ICs) are generally classified as digital (e.g. a microprocessors) or analog (e.g. an operational amplifier). Mixed-signal ICs are chips that contain both digital and analog circuits on the same chip. This category of chip has grown dramatically with the increased use of 3G cell phones and other portable technologies.

Mixed-signal ICs are often used to convert analog signals to digital signals so that digital devices can process them. For example, mixed-signal ICs are essential components for FM tuners in digital products such as media players, which have digital amplifiers. Any analog signal (such as an FM radio transmission, a light wave or a sound) can be digitized using a very basic analog-to-digital converter, and the smallest and most energy efficient of these would be in the form of mixed-signal ICs.

Mixed-signal ICs are more difficult to design and manufacture than analog-only or digital-only integrated circuits. For example, an efficient mixed-signal IC would have its digital and analog components share a common power supply. However, analog and digital components have very different power needs and consumption characteristics that make this a non-trivial goal in chip design.

Examples

Typically, mixed-signal chips perform some whole function or sub-function in a larger assembly such as the radio subsystem of a cell phone, or the read data path and laser sled control logic of a DVD player. They often contain an entire system-on-a-chip.

Examples of mixed-signal integrated circuits include data converters using delta-sigma modulation, analog-to-digital converter/digital-to-analog converter using error detection and correction, and digital radio chips. Digitally controlled sound chips are also mixed-signal circuits. With the advent of cellular technology and network technology this category now includes cellular telephone, software radio, LAN and WAN router integrated circuits.

Because of the use of both digital signal processing and analog circuitry, mixed-signal ICs are usually designed for a very specific purpose and their design requires a high level of expertise and careful use of computer aided design (CAD) tools. Automated testing of the finished chips can also be challenging. Teradyne, Keysight, and Texas Instruments are the major suppliers of the test equipment for mixed-signal chips.

The particular challenges of mixed signal include:

- CMOS technology is usually optimal for digital performance and scaling while bipolar transistors are usually optimal for analog performance, yet until the last decade it has been difficult to either combine these cost-effectively or to design both analog and digital in a single technology without serious performance compromises. The advent of technologies like high performance CMOS, BiCMOS, CMOS SOI and SiGe have removed many of the compromises that previously had to be made.

- Testing functional operation of mixed-signal ICs remains complex, expensive and often a "one-off" implementation task.

- Systematic design methodologies comparable to digital design methods are far more primitive in the analog and mixed-signal arena. Analog circuit design can not generally be automated to nearly the extent that digital circuit design can. Combining the two technologies multiplies this complication.

- Fast-changing digital signals send noise to sensitive analog inputs. One path for this noise is substrate coupling. A variety of techniques are used to attempt to block or cancel this noise coupling, such as fully differential amplifiers, P+ guard-rings, differential topology, on-chip decoupling, and triple-well isolation.

Commercial Examples

- ICsense
- AnSem
- Atari POKEY
- MOS Technology SID
- PSoC - Cypress PSoC Programmable System on Chip
- System to ASIC
- Texas Instruments' MSP430
- Triad Semiconductor
- Wolfson Microelectronics

Most modern radio and communications use mixed signal circuits.

Op Amp Integrator

The operational amplifier integrator is an electronic integration circuit. Based on the operational amplifier (op-amp), it performs the mathematical operation of integration with respect to time; that is, its output voltage is proportional to the input voltage integrated over time.

Applications

The integrator circuit is mostly used in analog computers, analog-to-digital converters and wave-shaping circuits. A common wave-shaping use is as a charge amplifier and they are usually constructed using an operational amplifier though they can use high gain discrete transistor configurations.

Design

The input current is offset by a negative feedback current flowing in the capacitor, which is generated by an increase in output voltage of the amplifier. The output voltage is therefore dependent on the value of input current it has to offset and the inverse of the value of the feedback capacitor. The greater the capacitor value, the less output voltage has to be generated to produce a particular feedback current flow.

The input impedance of the circuit is almost zero because of the Miller effect. Hence all the stray capacitances (the cable capacitance, the amplifier input capacitance, etc.) are virtually grounded and they have no influence on the output signal.

Ideal Circuit

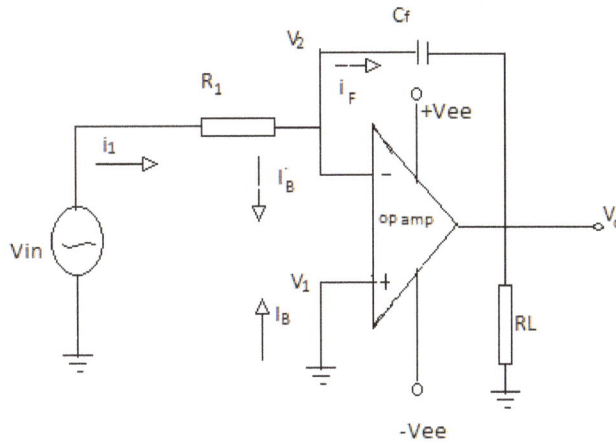

The circuit operates by passing a current that charges or discharges the capacitor C_f during the time under consideration, which strives to retain the virtual ground condition at the input by off-setting the effect of the input current. Referring to the above diagram, if the op-amp is assumed to be ideal, nodes v_1 and v_2 are held equal, and so v_2 is a virtual ground. The input voltage passes a current $\dfrac{v_{in}}{R_1}$ through the resistor producing a compensating current flow through the series capacitor to maintain the virtual ground. This charges or discharges the capacitor over time. Because the resistor and capacitor are connected to a virtual ground, the input current does not vary with capacitor charge and a linear integration of output is achieved.

The circuit can be analyzed by applying Kirchhoff's current law at the node v_2, keeping ideal op-amp behaviour in mind.

$$i_1 = I_B + i_F$$

$I_B = 0$ in an ideal op-amp, so:

$$i_1 = i_F$$

Furthermore, the capacitor has a voltage-current relationship governed by the equation:

$$I_C = C\frac{dV_c}{dt}$$

Substituting the appropriate variables:

$$\frac{v_{in} - v_2}{R_1} = C_F \frac{d(v_2 - v_o)}{dt}$$

$v_2 = v_1 = 0$ in an ideal op-amp, resulting in:

$$\frac{v_{in}}{R_1} = -C_F \frac{dv_o}{dt}$$

Integrating both sides with respect to time:

$$\int_0^t \frac{v_{in}}{R_1} dt = -\int_0^t C_F \frac{dv_o}{dt} dt$$

If the initial value of v_o is assumed to be 0 V, this results in a DC error of:

$$v_o = -\frac{1}{R_1 C_F} \int_0^t v_{in} dt$$

Practical Circuit

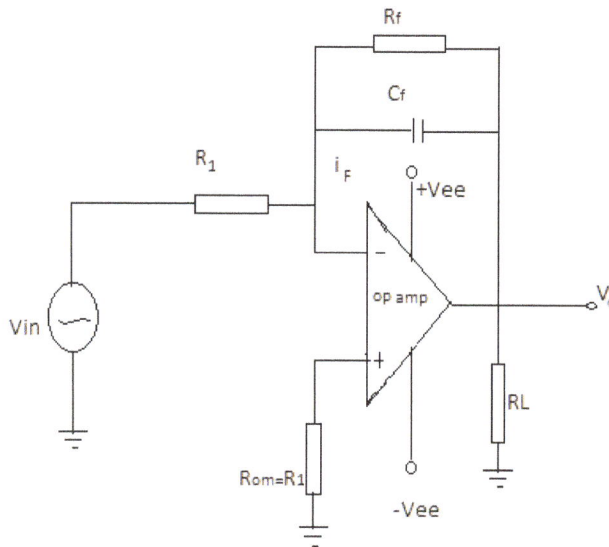

The ideal circuit is not a practical integrator design for a number of reasons. Practical op-amps have a finite open-loop gain, an input offset voltage and input bias currents (I_B). This can cause several issues for the ideal design; most importantly, if $v_{in} = 0$, both the output offset voltage and the input bias current I_B can cause current to pass through the capacitor, causing the output voltage to drift over time until the op-amp saturates. Similarly, if v_{in} were a signal centered about zero volts (i.e. without a DC component), no drift would be expected in an ideal circuit, but may occur in a real circuit. To negate the effect of the input bias current, it is necessary to set:

$R_{on} = R_1 \| R_f \| R_L$. The error voltage then becomes:

$$V_E = \left(\frac{R_f}{R_1} + 1\right) V_{IOS}$$

The input bias current thus causes the same voltage drops at both the positive and negative terminals.

Also, in a DC steady state, the capacitor acts as an open circuit. The DC gain of the ideal circuit is therefore infinite (or in practice, the open-loop gain of a non-ideal op-amp). To counter this, a large resistor R_F is inserted in parallel with the feedback capacitor, as shown in the figure above. This limits the DC gain of the circuit to a finite value, and hence changes the output drift into a finite, preferably small, DC error. Referring to the above diagram:

$$V_E = \left(\frac{R_f}{R_1} + 1\right)\left(V_{IOS} + I_{BI}\left(R_f \parallel R_1\right)\right)$$

where V_{IOS} is the input offset voltage and I_{BI} is the input bias current on the inverting terminal. $R_f \parallel R_1$ indicates two resistance values in parallel.

Frequency Response

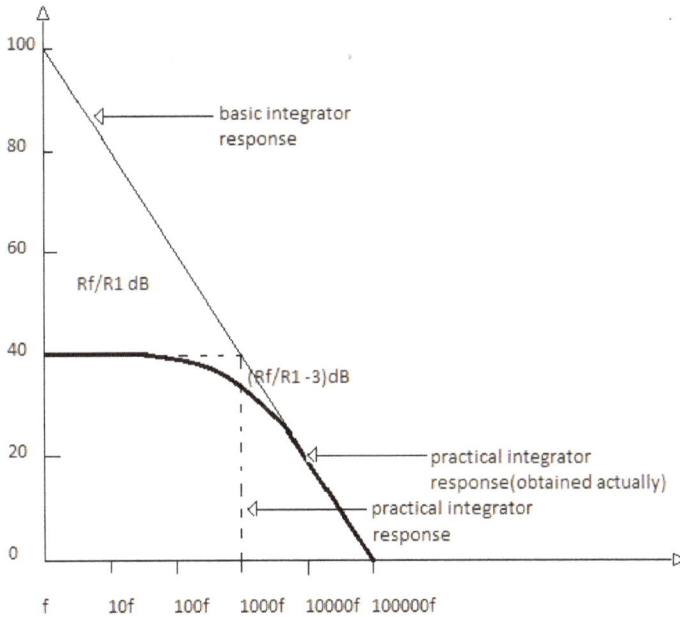

The frequency responses of the practical and ideal integrator are shown in the above figure. For both circuits, the crossover frequency f_b, at which the gain is 0 dB, is given by:

$$f_b = \frac{1}{2\pi R_1 C_F}$$

The 3 dB cutoff frequency f_a of the practical circuit is given by:

$$f_a = \frac{1}{2\pi R_F C_F}$$

The practical integrator circuit is equivalent to an active first-order low-pass filter. The gain is relatively constant up to the cutoff frequency and decreases by 20 dB per decade beyond it. The integration operation occurs for frequencies in the range $[f_a, f_b]$, provided that $f_a < f_b$. This condition can be achieved by appropriate choice of $R_F C_F$ and $R_1 C_F$ time constants.

Integrated Circuit Layout Design Protection

Layout designs (topographies) of integrated circuits are a field in the protection of intellectual property.

In United States intellectual property law, a "mask work" is a two or three-dimensional layout or topography of an integrated circuit (IC or "chip"), i.e. the arrangement on a chip of semiconductor devices such as transistors and passive electronic components such as resistors and interconnections. The layout is called a *mask* work because, in photolithographic processes, the multiple etched layers within actual ICs are each created using a mask, called the photomask, to permit or block the light at specific locations, sometimes for hundreds of chips on a wafer simultaneously.

Because of the functional nature of the mask geometry, the designs cannot be effectively protected under copyright law (except perhaps as decorative art). Similarly, because individual lithographic mask works are not clearly protectable subject matter, they also cannot be effectively protected under patent law, although any processes implemented in the work may be patentable. So since the 1990s, national governments have been granting copyright-like exclusive rights conferring time-limited exclusivity to reproduction of a particular layout.

International Law

A diplomatic conference was held at Washington, D.C., in 1989, which adopted a Treaty on Intellectual Property in Respect of Integrated Circuits, also called the Washington Treaty or IPIC Treaty. The Treaty, signed at Washington on May 26, 1989, is open to States Members of WIPO or the United Nations and to intergovernmental organizations meeting certain criteria. The Treaty has been incorporated by reference into the TRIPS Agreement of the World Trade Organization (WTO), subject to the following modifications: the term of protection is at least 10 (rather than eight) years from the date of filing an application or of the first commercial exploitation in the world, but Members may provide a term of protection of 15 years from the creation of the layout-design; the exclusive right of the right-holder extends also to articles incorporating integrated circuits in which a protected layout-design is incorporated, in so far as it continues to contain an unlawfully reproduced layout-design; the circumstances in which layout-designs may be used without the consent of right-holders are more restricted; certain acts engaged in unknowingly will not constitute infringement.

The IPIC Treaty is currently not in force, but was partially integrated into the TRIPS agreement.

Article 35 of TRIPS in Relation to the IPIC Treaty states:

Members agree to provide protection to the layout-designs (topographies) of integrated circuits (referred to in this Agreement as "layout-designs") in accordance with Articles 2 through 7 (other than paragraph 3 of Article 6), Article 12 and paragraph 3 of Article 16 of the Treaty on Intellectual Property in Respect of Integrated Circuits and, in addition, to comply with the following provisions. TRIPS Document

Article 2 of the IPIC Treaty gives the following definitions:

(i) 'integrated circuit' means a product, in its final form or an intermediate form, in which the elements, at least one of which is an active element, and some or all of the inter-connections are integrally formed in and/or on a piece of material and which is intended to perform an electronic function,

(ii) 'layout-design (topography)' means the three-dimensional disposition, however expressed, of the elements, at least one of which is an active element, and of some or all of the interconnections of an integrated circuit, or such a three-dimensional disposition prepared for an integrated circuit intended for manufacture ...

Under the IPIC Treaty, each Contracting Party is obliged to secure, throughout its territory, exclusive rights in layout-designs (topographies) of integrated circuits, whether or not the integrated circuit concerned is incorporated in an article. Such obligation applies to layout-designs that are original in the sense that they are the result of their creators' own intellectual effort and are not commonplace among creators of layout designs and manufacturers of integrated circuits at the time of their creation.

The Contracting Parties must, as a minimum, consider the following acts to be unlawful if performed without the authorization of the holder of the right: the reproduction of the lay-out design, and the importation, sale or other distribution for commercial purposes of the layout-design or an integrated circuit in which the layout-design is incorporated. However, certain acts may be freely performed for private purposes or for the sole purpose of evaluation, analysis, research or teaching.

National Laws

United States

The United States Code (USC) defines a mask work as "a series of related images, however fixed or encoded, having or representing the predetermined, three-dimensional pattern of metallic, insulating, or semiconductor material present or removed from the layers of a semiconductor chip product, and in which the relation of the images to one another is such that each image has the pattern of the surface of one form of the semiconductor chip product" [(17 U.S.C. § 901(a)(2))]. Mask work exclusive rights were first granted in the US by the Semiconductor Chip Protection Act of 1984.

According to 17 U.S.C. § 904, rights in semiconductor mask works last 10 years. This contrasts with a term of 95 years for modern copyrighted works with a corporate authorship; alleged infringement of mask work rights are also not protected by a statutory fair use defense, nor by the typical backup copy exemptions that 17 U.S.C. § 117 provides for computer software. Nevertheless, as fair use in copyrighted works was originally recognized by the judiciary over a century before being codified in the Copyright Act of 1976, it is possible that the courts might likewise find a similar defense applies to mask work.

The non-obligatory symbol used in a mask work protection notice is Ⓜ (M enclosed in a circle; Unicode code point U+24C2 or HTML numeric character entity Ⓜ) or *M*.

The exclusive rights in a mask work are somewhat like those of copyright: the right to reproduce the mask work or (initially) distribute an IC made using the mask work. Like the first sale doctrine, a lawful owner of an authorized IC containing a mask work may freely import, distribute or use, but not reproduce the chip (or the mask). Mask work protection is characterized as a *sui generis* right, i.e., one created to protect specific rights where other (more general) laws were inadequate or inappropriate.

Note that the exclusive rights granted to mask work owners are more limited than those granted to copyright or patent holders. For instance, modification (derivative works) is not an exclusive right of mask work owners. Similarly, the exclusive right of a patentee to "use" an invention would not prohibit an independently created mask work of identical geometry. Furthermore, reproduction for reverse engineering of a mask work is specifically permitted by the law. As with copyright, mask work rights exist when they are created, regardless of registration, unlike patents, which only confer rights after application, examination and issuance.

Mask work rights have more in common with copyrights than with other exclusive rights such as patents or trademarks. On the other hand, they are used alongside copyright to protect a read-only memory (ROM) component that is encoded to contain computer software.

The publisher of software for a cartridge-based video game console may seek simultaneous protection of its property under several legal constructs:

- A trademark registration on the game's title and possibly other marks such as fanciful names of worlds and characters used in the game (e.g., PAC-MAN®);

- A copyright registration on the program as a literary work or on the audiovisual displays generated by the work; and

- A mask work registration on the ROM that contains the binary.

Ordinary copyright law applies to the underlying software (source, binary) and original characters and art. But the expiration date for the term of *additional* exclusive rights in a work distributed in the form of a mask ROM would depend on an as yet untested interpretation of the originality requirement of § 902(b):

(b) Protection under this chapter (i.e., as a mask work) shall not be available for a mask work that—

   (1) is not original; or

   (2) consists of designs that are staple, commonplace, or familiar in the semiconductor industry, or variations of such designs, combined in a way that, considered as a whole, is not original

(17 U.S.C. § 902, as of November 2010).

Under one interpretation, a mask work containing a given game title is either entirely unoriginal, as mask ROM in general is likely a familiar design, or a minor variation of the mask work for any of the first titles released for the console in the region.

Other Countries

Equivalent legislation exists in Australia, India and Hong Kong.

In Canada these rights are protected under the [Integrated Circuit Topography Act (1990, c. 37)].

In the European Union, a *sui generis* design right protecting the design of materials was introduced by the Directive 87/54/EEC which is transposed in all member states.

India has the Semiconductor Integrated Circuits Layout Design Act, 2000 for the similar protection.

Japan relies on the "The Act Concerning the Circuit Layout of a Semiconductor Integrated Circuit".

Brazil has enacted Law No. 11484, of 2007, to regulate the protection and registration of integrated circuit topography.

References

- Paul R. Gray, Paul J. Hurst, Stephen H. Lewis, and Robert G. Meyer (2009). Analysis and Design of Analog Integrated Circuits. Wiley. ISBN 978-0470245996.

- Jan M. Rabaey, Anantha Chandrakasan, and Borivoje Nikolic (2003). Digital Integrated Circuits (2nd Edition). Pearson. ISBN 978-0130909961.

- Winston, Brian (1998). Media Technology and Society: A History : From the Telegraph to the Internet. Routledge. p. 221. ISBN 978-0-415-14230-4.

- Ginzberg, Eli (1976). Economic impact of large public programs: the NASA Experience. Olympus Publishing Company. p. 57. ISBN 0-913420-68-9.

- Bob Johnstone (1999). We were burning: Japanese entrepreneurs and the forging of the electronic age. Basic Books. pp. 47–48. ISBN 978-0-465-09118-8.

- Braun, E.; MacDonald, S. (1982). Revolution in Miniature: The History and Impact of Semiconductor Electronics (2nd ed.). Cambridge University Press. ISBN 9780521289030.

- Bassett, R. K. (2007). "RCA and the Quest for Radical Technological Change". To the Digital Age: Research Labs, Start-Up Companies, and the Rise of MOS Technology. JHU Press. ISBN 9780801886393.

- Morris, P. R. (1990). A history of the world semiconductor industry. History of technology series. 12. IET. pp. 34, 36. ISBN 9780863412271.

- Seitz, F.; Einspruch, N. (1998). Electronic genie: the tangled history of silicon. University of Illinois Press. p. 214. ISBN 9780252023835.

- Swain, P.; Gill, J. (1993). Corporate Vision and Rapid Technological Change: The Evolution of Market Structure. Routledge. pp. 140–143. ISBN 9780415091350.

- Reid, T. R. (1984). The Chip: How Two Americans Invented the Microchip and Launched a Revolution. Simon and Schuster. ISBN 9780671453930.

- Reid, T. (1984). The Chip: How Two Americans Invented the Microchip and Launched a Revolution. Simon and Schuster. p. 76. ISBN 9780671453930.

- Greig, William (2007). Integrated Circuit Packaging, Assembly and Interconnections. Springer Science &

Business Media. ISBN 9780387339139.

- Baker, R. Jacob (2010). CMOS: Circuit Design, Layout, and Simulation, Third Edition. Wiley-IEEE. ISBN 978-0-470-88132-3.

- Ken Gilleo (2003). Area array packaging processes for BGA, Flip Chip, and CSP. McGraw-Hill Professional. p. 251. ISBN 0-07-142829-1.

3

Linear Circuit: An Integrated Study

Linear circuits are electronic circuits; a better definition of linear circuits that can be given is that it follows the superposition principle. The section strategically encompasses and incorporates the major components and key concepts of linear circuit, providing a complete understanding.

Linear circuit

A linear circuit is an electronic circuit in which, for a sinusoidal input voltage of frequency f, any steady-state output of the circuit (the current through any component, or the voltage between any two points) is also sinusoidal with frequency f. Note that the output need not be in phase with the input.

An equivalent definition of a linear circuit is that it obeys the superposition principle. This means that the output of the circuit $F(x)$ when a linear combination of signals $ax_1(t) + bx_2(t)$ is applied to it is equal to the linear combination of the outputs due to the signals $x_1(t)$ and $x_2(t)$ applied separately:

$$F(ax_1 + bx_2) = aF(x_1) + bF(x_2)$$

It is called a linear circuit because the output of such a circuit is a linear function of its inputs. Informally, a linear circuit is one in which the electronic components' values (such as resistance, capacitance, inductance, gain, etc.) do not change with the level of voltage or current in the circuit. Linear circuits are important because they can amplify and process electronic signals without distortion. An example of an electronic device that uses linear circuits is a sound system.

Linear and Nonlinear Components

A linear circuit is one that has no nonlinear electronic components in it. Examples of linear circuits are amplifiers, differentiators, and integrators, linear electronic filters, or any circuit composed exclusively of *ideal* resistors, capacitors, inductors, op-amps (in the "non-saturated" region), and other "linear" circuit elements.

Some examples of nonlinear electronic components are: diodes, transistors, and iron core inductors and transformers when the core is saturated. Some examples of circuits that operate in a nonlinear way are mixers, modulators, rectifiers, radio receiver detectors and digital logic circuits.

Significance

Linear circuits are important because they can process analog signals without introducing intermodulation distortion. This means that separate frequencies in the signal stay separate and do not mix, creating new frequencies (heterodynes).

They are also easier to understand. Because they obey the superposition principle, linear circuits can be analyzed with powerful mathematical frequency domain techniques, including Fourier analysis and the Laplace transform. These also give an intuitive understanding of the qualitative behavior of the circuit, characterizing it using terms such as gain, phase shift, resonant frequency, bandwidth, Q factor, poles, and zeros. The analysis of a linear circuit can often be done by hand using a scientific calculator.

In contrast, nonlinear circuits usually do not have closed form solutions. They must be analyzed using approximate numerical methods by electronic circuit simulation computer programs such as SPICE, if accurate results are desired. The behavior of such linear circuit elements as resistors, capacitors, and inductors can be specified by a single number (resistance, capacitance, inductance, respectively). In contrast, a nonlinear element's behavior is specified by its detailed transfer function, which may be given as a graph. So specifying the characteristics of a nonlinear circuit requires more information than is needed for a linear circuit.

"Linear" circuits and systems form a separate category within electronic manufacturing. Manufacturers of transistors and integrated circuits often divide their product lines into 'linear' and 'digital' lines. "Linear" here means "analog"; the linear line includes integrated circuits designed to process signals linearly, such as op-amps, audio amplifiers, and active filters, as well as a variety of signal processing circuits that implement nonlinear analog functions such as logarithmic amplifiers, analog multipliers, and peak detectors.

Small Signal Approximation

Nonlinear elements such as transistors tend to behave linearly when small AC signals are applied to them. So in analysing many circuits where the signal levels are small, for example those in TV and radio receivers, nonlinear elements can be replaced with a linear small-signal model, allowing linear analysis techniques to be used.

Conversely, all circuit elements, even "linear" elements, show nonlinearity as the signal level is increased. If nothing else, the power supply voltage to the circuit usually puts a limit on the magnitude of voltage output from a circuit. Above that limit, the output ceases to scale in magnitude with the input, failing the definition of linearity.

LC Circuit

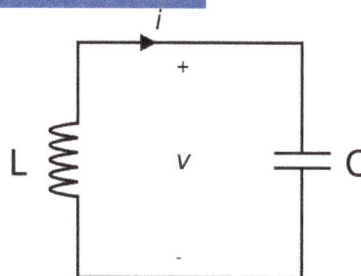

LC circuit diagram

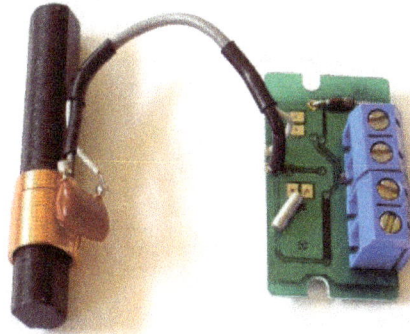

LC circuit *(left)* consisting of ferrite coil and capacitor used as a tuned circuit in the receiver for a radio clock.

An LC circuit, also called a resonant circuit, tank circuit, or tuned circuit, is an electric circuit consisting of an inductor, represented by the letter L, and a capacitor, represented by the letter C, connected together. The circuit can act as an electrical resonator, an electrical analogue of a tuning fork, storing energy oscillating at the circuit's resonant frequency.

LC circuits are used either for generating signals at a particular frequency, or picking out a signal at a particular frequency from a more complex signal. They are key components in many electronic devices, particularly radio equipment, used in circuits such as oscillators, filters, tuners and frequency mixers.

An LC circuit is an idealized model since it assumes there is no dissipation of energy due to resistance. Any practical implementation of an LC circuit will always include loss resulting from small but non-zero resistance within the components and connecting wires. The purpose of an LC circuit is usually to oscillate with minimal damping, so the resistance is made as low as possible. While no practical circuit is without losses, it is nonetheless instructive to study this ideal form of the circuit to gain understanding and physical intuition. For a circuit model incorporating resistance.

Terminology

The two-element LC circuit described above is the simplest type of inductor-capacitor network (or LC network). It is also referred to as a *second order LC circuit* to distinguish it from more complicated (higher order) LC networks with more inductors and capacitors. Such LC networks with more than two reactances may have more than one resonant frequency.

The order of the network is the order of the rational function describing the network in the complex frequency variable s. Generally, the order is equal to the number of L and C elements in the circuit and in any event cannot exceed this number.

Operation

An LC circuit, oscillating at its natural resonant frequency, can store electrical energy. A capacitor stores energy in the electric ield (E) between its plates, depending on the voltage across it, and an inductor stores energy in its magnetic ield (B), depending on the current through it.

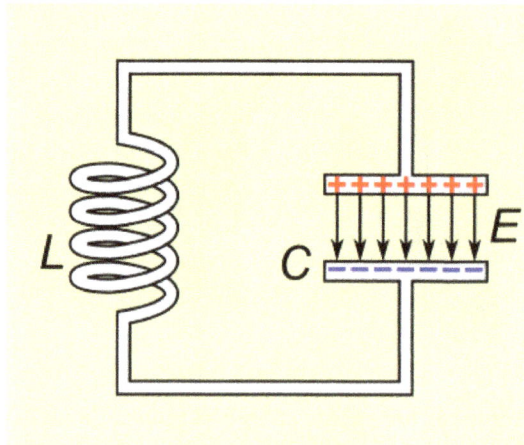

Animated diagram showing the operation of a tuned circuit (LC circuit). The capacitor C stores energy in its electric field E and the inductor L stores energy in its magnetic field B *(green)*. This jerky animation shows "snapshots" of the circuit at progressive points in the oscillation. The oscillations are slowed down; in an actual tuned circuit the charge oscillates back and forth tens of thousands to billions of times per second.

If an inductor is connected across a charged capacitor, current will start to flow through the inductor, building up a magnetic field around it and reducing the voltage on the capacitor. Eventually all the charge on the capacitor will be gone and the voltage across it will reach zero. However, the current will continue, because inductors resist changes in current. The current will begin to charge the capacitor with a voltage of opposite polarity to its original charge. Due to Faraday's law, the EMF which drives the current is caused by a decrease in the magnetic field, thus the energy required to charge the capacitor is extracted from the magnetic field. When the magnetic field is completely dissipated the current will stop and the charge will again be stored in the capacitor, with the opposite polarity as before. Then the cycle will begin again, with the current flowing in the opposite direction through the inductor.

The charge flows back and forth between the plates of the capacitor, through the inductor. The energy oscillates back and forth between the capacitor and the inductor until (if not replenished from an external circuit) internal resistance makes the oscillations die out. In most applications the tuned circuit is part of a larger circuit which applies alternating current to it, driving continuous oscillations. If these are at the natural oscillatory frequency (Natural frequency), resonance will occur. The tuned circuit's action, known mathematically as a harmonic oscillator, is similar to a pendulum swinging back and forth, or water sloshing back and forth in a tank; for this reason the circuit is also called a *tank circuit*. The natural frequency (that is, the frequency at which it will oscillate when isolated from any other system, as described above) is determined by the capacitance and inductance values. In typical tuned circuits in electronic equipment the oscillations are very fast, thousands to billions of times per second.

Resonance Effect

Resonance occurs when an LC circuit is driven from an external source at an angular frequency ω_0 at which the inductive and capacitive reactances are equal in magnitude. The frequency at which this equality holds for the particular circuit is called the resonant frequency. The resonant frequency of the LC circuit is

$$\omega_0 = \frac{1}{\sqrt{LC}}$$

where L is the inductance in henrys, and C is the capacitance in farads. The angular frequency ω_0 has units of radians per second.

The equivalent frequency in units of hertz is

$$f_0 = \frac{\omega_0}{2\pi} = \frac{1}{2\pi\sqrt{LC}}.$$

LC circuits are often used as filters; the L/C ratio is one of the factors that determines their "Q" and so selectivity. For a series resonant circuit with a given resistance, the higher the inductance and the lower the capacitance, the narrower the filter bandwidth. For a parallel resonant circuit the opposite applies. Positive feedback around the tuned circuit ("regeneration") can also increase selectivity.

Stagger tuning can provide an acceptably wide audio bandwidth, yet good selectivity.

Applications

The resonance effect of the LC circuit has many important applications in signal processing and communications systems.

- The most common application of tank circuits is tuning radio transmitters and receivers. For example, when we tune a radio to a particular station, the LC circuits are set at resonance for that particular carrier frequency.

- A series resonant circuit provides voltage magnification.

- A parallel resonant circuit provides current magnification.

- A parallel resonant circuit can be used as load impedance in output circuits of RF amplifiers. Due to high impedance, the gain of amplifier is maximum at resonant frequency.

- Both parallel and series resonant circuits are used in induction heating.

LC circuits behave as electronic resonators, which are a key component in many applications:

- Amplifiers

- Oscillators

- Filters

- Tuners

- Mixers

- Foster-Seeley discriminator

- Contactless cards

- Graphics tablets

- Electronic article surveillance (security tags)

Time Domain Solution

Kirchhoff's Laws

By Kirchhoff's voltage law, the voltage across the capacitor, V_C, plus the voltage across the inductor, V_L must equal zero:

$$V_C + V_L = 0.$$

Likewise, by Kirchhoff's current law, the current through the capacitor equals the current through the inductor:

$$I_C = I_L.$$

From the constitutive relations for the circuit elements, we also know that

$$V_L(t) = L\frac{dI_L}{dt},$$

$$I_C(t) = C\frac{dV_C}{dt}.$$

Differential Equation

Rearranging and substituting gives the second order differential equation

$$\frac{d^2}{dt^2}I(t) + \frac{1}{LC}I(t) = 0.$$

The parameter ω_0, the resonant angular frequency, is defined as:

$$\omega_0 = \frac{1}{\sqrt{LC}}.$$

Using this can simplify the differential equation

$$\frac{d^2}{dt^2}I(t) + \omega_0^2 I(t) = 0.$$

The associated polynomial is

$$s^2 + \omega_0^2 = 0;$$

thus,

$$s = \pm j\omega_0,$$

where j is the imaginary unit.

Solution

Thus, the complete solution to the differential equation is

$$I(t) = Ae^{+j\omega_0 t} + Be^{-j\omega_0 t}$$

and can be solved for A and B by considering the initial conditions. Since the exponential is complex, the solution represents a sinusoidal alternating current. Since the electric current I is a physical quantity, it must be real-valued. As a result, it can be shown that the constants A and B must be complex conjugates:

$$A = B^*$$

Now, let

$$A = \frac{I_0}{2}e^{+j\phi}$$

Therefore,

$$B = \frac{I_0}{2}e^{-j\phi}$$

Next, we can use Euler's formula to obtain a real sinusoid with amplitude I_0, angular frequency ω_0 = 1/√LC, and phase angle φ.

Thus, the resulting solution becomes:

$$I(t) = I_0 \cos(\omega_0 t + \phi).$$

and

$$V(t) = L\frac{dI}{dt} = -\omega_0 LI_0 \sin(\omega_0 t + \phi).$$

Initial Conditions

The initial conditions that would satisfy this result are:

$$I(0) = I_0 \cos\phi.$$

and

$$V(0) = L\frac{dI}{dt}\Big|_{t=0} = -\omega_0 LI_0 \sin\phi.$$

Series LC Circuit

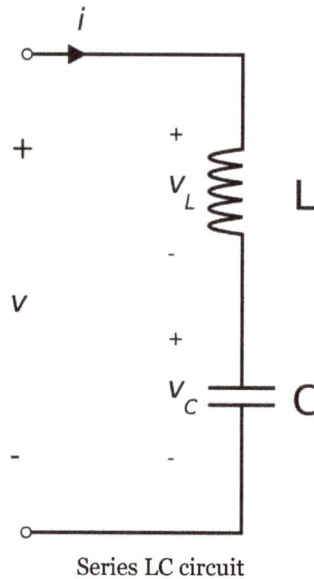

Series LC circuit

In the series configuration of the LC circuit, the inductor (L) and capacitor (C) are connected in series, as shown here. The total voltage V across the open terminals is simply the sum of the voltage across the inductor and the voltage across the capacitor. The current I into the positive terminal of the circuit is equal to the current through both the capacitor and the inductor.

$$V = V_L + V_C$$
$$I = I_L = I_C.$$

Resonance

Inductive reactance magnitude X_L increases as frequency increases while capacitive reactance magnitude X_C decreases with the increase in frequency. At one particular frequency, these two reactances are equal in magnitude but opposite in sign; that frequency is called the resonant frequency f_o for the given circuit.

Hence, at resonance:

$$X_L = -X_C \omega$$
$$L = \frac{1}{\omega C}.$$

Solving for ω, we have

$$\omega = \omega_0 = \frac{1}{\sqrt{LC}},$$

which is defined as the resonant angular frequency of the circuit. Converting angular frequency (in radians per second) into frequency (in hertz), one has

$$f_0 = \frac{\omega_0}{2\pi} = \frac{1}{2\pi\sqrt{LC}}.$$

In a series configuration, X_C and X_L cancel each other out. In real, rather than idealised components, the current is opposed, mostly by the resistance of the coil windings. Thus, the current supplied to a series resonant circuit is a maximum at resonance.

- In the limit as $f \to f_0$ current is maximum. Circuit impedance is minimum. In this state, a circuit is called an *acceptor circuit*.

- For $f < f_0$, $X_L \ll -X_C$. Hence, the circuit is capacitive.

- For $f > f_0$, $X_L \gg -X_C$. Hence, the circuit is inductive.

Impedance

In the series configuration, resonance occurs when the complex electrical impedance of the circuit approaches zero.

First consider the impedance of the series LC circuit. The total impedance is given by the sum of the inductive and capacitive impedances:

$$Z = Z_L + Z_C$$

Writing the inductive impedance as $Z_L = j\omega L$ and capacitive impedance as $Z_C = 1/j\omega C$ and substituting gives

$$Z(\omega) = j\omega L + \frac{1}{j\omega C}.$$

Writing this expression under a common denominator gives

$$Z(\omega) = j\left(\frac{\omega^2 LC - 1}{\omega C}\right).$$

Finally, defining the natural angular frequency as

$$\omega_0 = \frac{1}{\sqrt{LC}},$$

the impedance becomes

$$Z(\omega) = jL\left(\frac{\omega^2 - \omega_0^2}{\omega}\right).$$

The numerator implies that in the limit as $\omega \to \pm\omega_0$, the total impedance Z will be zero and otherwise non-zero. Therefore the series LC circuit, when connected in series with a load, will act as a band-pass filter having zero impedance at the resonant frequency of the LC circuit.

Parallel LC Circuit

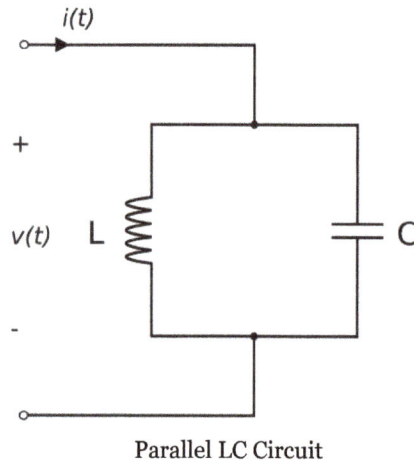

Parallel LC Circuit

In the parallel configuration, the inductor L and capacitor C are connected in parallel, as shown here. The voltage V across the open terminals is equal to both the voltage across the inductor and the voltage across the capacitor. The total current I flowing into the positive terminal of the circuit is equal to the sum of the current flowing through the inductor and the current flowing through the capacitor:

$$V = V_L = V_C$$
$$I = I_L + I_C.$$

Resonance

When X_L equals X_C, the reactive branch currents are equal and opposite. Hence they cancel out each other to give minimum current in the main line. Since total current is minimum, in this state the total impedance is maximum.

The resonant frequency is given by

$$f_0 = \frac{\omega_0}{2\pi} = \frac{1}{2\pi\sqrt{LC}}.$$

Note that any reactive branch current is not minimum at resonance, but each is given separately by dividing source voltage (V) by reactance (Z). Hence $I = V/Z$, as per Ohm's law.

- At f_0, the line current is minimum. The total impedance is at the maximum. In this state a circuit is called a *rejector circuit*.

- Below f_0, the circuit is inductive.

- Above f_0, the circuit is capacitive.

Impedance

The same analysis may be applied to the parallel LC circuit. The total impedance is then given by:

$$Z = \frac{Z_L Z_C}{Z_L + Z_C},$$

and after substitution of Z_L and Z_C and simplification, gives

$$Z(\omega) = -j \cdot \frac{\omega L}{\omega^2 LC - 1}$$

which further simplifies to

$$Z(\omega) = -j \left(\frac{1}{C} \right) \left(\frac{\omega}{\omega^2 - \omega_0^2} \right),$$

where

$$\omega_0 = \frac{1}{\sqrt{LC}}.$$

Note that

$$\lim_{\omega \to \pm \omega_0} Z(\omega) = \infty$$

but for all other values of ω the impedance is finite. The parallel LC circuit connected in series with a load will act as band-stop filter having infinite impedance at the resonant frequency of the LC circuit. The parallel LC circuit connected in parallel with a load will act as band-pass filter.

Laplace Solution

The LC circuit can be solved by Laplace transform.

Let the general equation be:

$$v_C(t) = v(t)$$

$$i(t) = C \frac{dv_C}{dt}$$

$$v_L(t) = L \frac{di}{dt}$$

Let the differential equation of LC series be:

$$v_{in}(t) = v_L(t) + v_C(t) = L\frac{di}{dt} + v = LC\frac{d^2v}{dt^2} + v$$

With initial condition:

$$\begin{cases} v(0) = v_0 \\ i(0) = i_0 = C * v'(0) = C * v'_0 \end{cases}$$

Let define:

$$\omega_0 = \frac{1}{\sqrt{LC}}$$

$$f(t) = \omega_0^2 v_{in}(t)$$

Gives:

$$f(t) = \frac{d^2v}{dt^2} + \omega_0^2 v$$

Transform with Laplace:

$$\mathcal{L}[f(t)] = \mathcal{L}\left[\frac{d^2v}{dt^2} + \omega_0^2 v\right]$$

$$F(s) = s^2 V(s) - s v_0 - v'_0 + \omega_0^2 V(s)$$

$$V(s) = \frac{s v_0 + v'_0 + F(s)}{s^2 + \omega_0^2}$$

Then antitransform:

$$v(t) = v_0 \cos(\omega_0 t) + \frac{v'_0}{\omega_0}\sin(\omega_0 t) + \mathcal{L}^{-1}\left[\frac{F(s)}{s^2 + \omega_0^2}\right]$$

In case input voltage is Heaviside step function:

$$v_{in}(t) = Mu(t)$$

$$\mathcal{L}^{-1}\left[\omega_0^2 \frac{V_{in}(s)}{s^2 + \omega_0^2}\right] = \mathcal{L}^{-1}\left[\omega_0^2 M \frac{1}{s(s^2 + \omega_0^2)}\right] = M\omega_0(1 - \cos(\omega_0 t))$$

$$v(t) = v_0 \cos(\omega_0 t) + \frac{v_0'}{\omega_0} \sin(\omega_0 t) + M\omega_0 (1 - \cos(\omega_0 t))$$

In case input voltage is sinusoidal function:

$$v_{in}(t) = U \sin(\omega_f t) \Rightarrow V_{in}(s) = \frac{U\omega_f}{s^2 + \omega_f^2}$$

$$\mathcal{L}^{-1}\left[\omega_0^2 \frac{1}{s^2 + \omega_0^2} \frac{U\omega_f}{s^2 + \omega_f^2}\right] = \mathcal{L}^{-1}\left[\frac{\omega_0^2 U\omega_f}{\omega_f^2 - \omega_0^2}\left(\frac{1}{s^2 + \omega_0^2} - \frac{1}{s^2 + \omega_f^2}\right)\right] = \frac{\omega_0^2 U\omega_f}{\omega_f^2 - \omega_0^2}\left(\frac{1}{\omega_0}\sin(\omega_0 t) - \frac{1}{\omega_f}\sin(\omega_f t)\right)$$

$$v(t) = v_0 \cos(\omega_0 t) + \frac{v_0'}{\omega_0}\sin(\omega_0 t) + \frac{\omega_0^2 U\omega_f}{\omega_f^2 - \omega_0^2}\left(\frac{1}{\omega_0}\sin(\omega_0 t) - \frac{1}{\omega_f}\sin(\omega_f t)\right)$$

History

The first evidence that a capacitor and inductor could produce electrical oscillations was discovered in 1826 by French scientist Felix Savary. He found that when a Leyden jar was discharged through a wire wound around an iron needle, sometimes the needle was left magnetized in one direction and sometimes in the opposite direction. He correctly deduced that this was caused by a damped oscillating discharge current in the wire, which reversed the magnetization of the needle back and forth until it was too small to have an effect, leaving the needle magnetized in a random direction. American physicist Joseph Henry repeated Savary's experiment in 1842 and came to the same conclusion, apparently independently. British scientist William Thomson (Lord Kelvin) in 1853 showed mathematically that the discharge of a Leyden jar through an inductance should be oscillatory, and derived its resonant frequency. British radio researcher Oliver Lodge, by discharging a large battery of Leyden jars through a long wire, created a tuned circuit with its resonant frequency in the audio range, which produced a musical tone from the spark when it was discharged. In 1857, German physicist Berend Wilhelm Feddersen photographed the spark produced by a resonant Leyden jar circuit in a rotating mirror, providing visible evidence of the oscillations. In 1868, Scottish physicist James Clerk Maxwell calculated the effect of applying an alternating current to a circuit with inductance and capacitance, showing that the response is maximum at the resonant frequency. The first example of an electrical resonance curve was published in 1887 by German physicist Heinrich Hertz in his pioneering paper on the discovery of radio waves, showing the length of spark obtainable from his spark-gap LC resonator detectors as a function of frequency.

One of the first demonstrations of resonance between tuned circuits was Lodge's "syntonic jars" experiment around 1889. He placed two resonant circuits next to each other, each consisting of a Leyden jar connected to an adjustable one-turn coil with a spark gap. When a high voltage from an induction coil was applied to one tuned circuit, creating sparks and thus oscillating currents, sparks were excited in the other tuned circuit only when the circuits were adjusted to resonance. Lodge and some English scientists preferred the term "*syntony*" for this effect, but the term "*resonance*" eventually stuck. The first practical use for LC circuits was in the 1890s in spark-gap radio transmitters to allow the receiver and transmitter to be tuned to the same frequency. The first patent for a radio system that allowed tuning was filed by Lodge in 1897, although the first practical systems were invented in 1900 by Italian radio pioneer Guglielmo Marconi.

Linear Amplifier

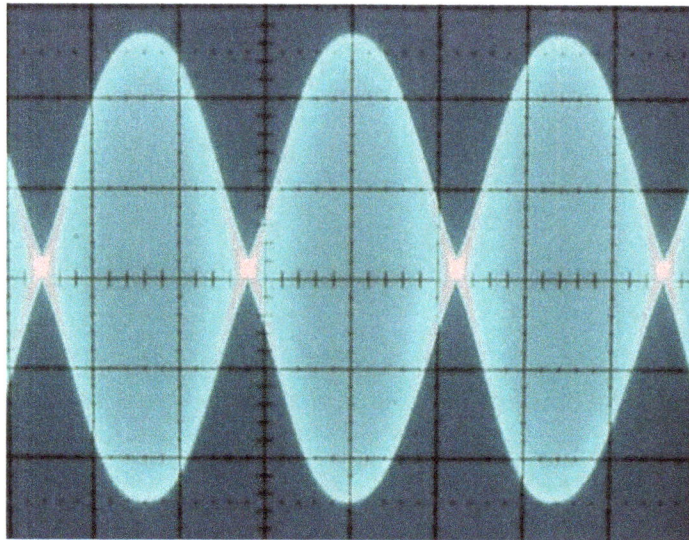

Linearity testing of a single-sideband transmitter

A linear amplifier is an electronic circuit whose output is proportional to its input, but capable of delivering more power into a load. The term usually refers to a type of radio-frequency (RF) power amplifier, some of which have output power measured in kilowatts, and are used in amateur radio. Other types of linear amplifier are used in audio and laboratory equipment.

Explanation

Linearity refers to the ability of the amplifier to produce signals that are accurate copies of the input, generally at increased power levels. Load impedance, supply voltage, input base current, and power output capabilities can affect the efficiency of the amplifier.

Class-A amplifiers can be designed to have good linearity in both *single ended* and *push-pull* topologies. Amplifiers of classes AB1, AB2 and B can be linear only in the push-pull topology, in which two active elements (tubes, transistors) are used to amplify positive and negative parts of the RF cycle respectively. Class-C amplifiers are not linear in any topology.

Amplifier Classes

There are a number of amplifier classes providing various trade-offs between implementation cost, efficiency, and signal accuracy. Their use in RF applications are listed briefly below:

- Class-A amplifiers are very inefficient, they can never have an efficiency better than 50%. The semiconductor or vacuum tube conducts throughout the entire RF cycle. The mean anode current for a vacuum tube should be set to the middle of the linear section of the curve of the anode current vs grid bias potential.

- Class B can be 60–65% efficient. The semiconductor or vacuum tube conducts through half the cycle but requires large drive power.

- Class AB1 is where the grid is more negatively biased than it is in class A.

- Class AB2 is where the grid is often more negatively biased than in AB1, also the size of the input signal is often larger. When the drive is able to make the grid become positive the grid current will increase.

- Class-C amplifiers are still more efficient. They can be about 75% efficient with a conduction range of about 120°, but they are very nonlinear. They can only be used for non-AM modes, such as FM, CW, or RTTY. The semiconductor or vacuum tube conducts through less than half the RF cycle. The increase in efficiency can allow a given vacuum tube to deliver more RF power than it could in class A or AB. For instance two 4CX250B tetrodes operating at 144 MHz can deliver 400 watts in class A, but when biased into class C they can deliver 1000 watts without fear of overheating. Even more grid current will be needed.

Although class-A power amplifiers (PA) are best in terms of linearity, their efficiency is rather poor as compared with other amplification classes such as "AB", "C" and Doherty amplifiers. However, higher efficiency leads to higher nonlinearity and PA output will be distorted, often to extent that fails the system performance requirements. Therefore, class-AB power amplifiers or other variations are used with some suitable form of linearization schemes such as feedback, feedforward or analog or digital predistortion (DPD). In DPD power amplifier systems, the transfer characteristics of the amplifier are modeled by sampling the output of the PA and the inverse characteristics are calculated in a DSP processor. The digital baseband signal is multiplied by the inverse of PA nonlinear transfer characteristics, up-converted to RF frequencies and is applied to the PA input. With careful design of PA response, the DPD engines can correct the PA output distortion and achieve higher efficiencies.

With advances in digital signal processing techniques, Digital Predistortion (DPD) is now widely used for RF power amplifier subsystems. In order for a DPD to function properly the power amplifier characteristics need to be optimal and circuit techniques are available to optimize the PA performance.

Amateur Radio

Most commercially manufactured one to two kilowatt linear amplifiers used in amateur radio still use vacuum tubes (valves) and can provide 10 to 20 times RF power amplification (10 to 13 dB). For example, a transmitter driving the input with 100 watts will be amplified to 2000 watts (2 kW) output to the antenna. Solid state linear amplifiers are more commonly in the 500 watt range and can be driven by as little as 25 watts.

Power triode Eimac 3CX1500A7

The maximum Amateur Radio Output is dependent on the licensed location, usually 1,500 to 2,250W. This is achieved, usually, with a linear amplifier. Large vacuum-tube linear amplifiers are based on old radio broadcast techniques and generally rely on a pair of large vacuum tubes supplied by a very high voltage power supply to convert large amounts of electrical energy into radio frequency energy. Linear amplifiers need to operate with class-A or class-AB biasing, which makes them relatively inefficient. While class C has far higher efficiency, a class-C amplifier is not linear, and is only suitable for the amplification of constant envelope signals. Such signals include FM, FSK, MFSK, and CW (Morse code).

Broadcast Radio Stations

The output stages of professional AM radio broadcast transmitters of up to 50 kW need to be linear and are now usually constructed using solid state technologies. Large vacuum tubes are still used for international long, medium, and shortwave broadcast transmitters from 500 kW up to 2 MW.

Thévenin's Theorem

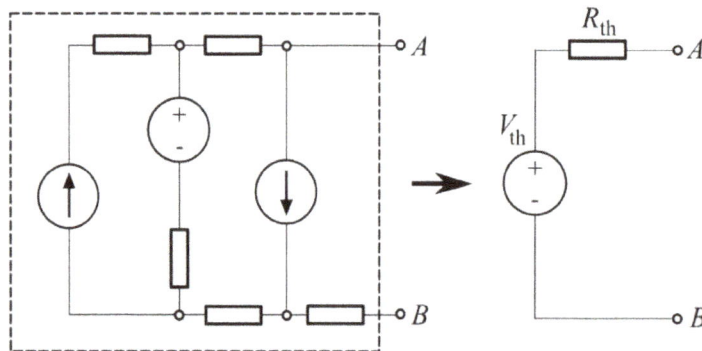

Any black box containing resistances only and voltage and current sources can be replaced by a Thévenin equivalent circuit consisting of an equivalent voltage source in series connection with an equivalent resistance.

As originally stated in terms of DC resistive circuits only, Thévenin's theorem holds that:

- Any linear electrical network with voltage and current sources and only resistances can be replaced at terminals A-B by an equivalent voltage source V_{th} in series connection with an equivalent resistance R_{th}.

- The equivalent voltage V_{th} is the voltage obtained at terminals A-B of the network with terminals A-B open circuited.

- The equivalent resistance R_{th} is the resistance that the circuit between terminals A and B would have if all ideal voltage sources in the circuit were replaced by a short circuit and all ideal current sources were replaced by an open circuit.

- If terminals A and B are connected to one another, the current flowing from A to B will be V_{th}/R_{th}. This means that R_{th} could alternatively be calculated as V_{th} divided by the short-circuit current between A and B when they are connected together.

In circuit theory terms, the theorem allows any one-port network to be reduced to a single voltage source and a single impedance.

The theorem also applies to frequency domain AC circuits consisting of reactive and resistive impedances.

The theorem was independently derived in 1853 by the German scientist Hermann von Helmholtz and in 1883 by Léon Charles Thévenin (1857–1926), an electrical engineer with France's national Postes et Télégraphes telecommunications organization.

Thévenin's theorem and its dual, Norton's theorem, are widely used to make circuit analysis simpler and to study a circuit's initial-condition and steady-state response. Thévenin's theorem can be used to convert any circuit's sources and impedances to a Thévenin equivalent; use of the theorem may in some cases be more convenient than use of Kirchhoff's circuit laws.

Calculating the Thévenin Equivalent

The equivalent circuit is a voltage source with voltage V_{Th} in series with a resistance R_{Th}.

The Thévenin-equivalent voltage V_{Th} is the voltage at the output terminals of the original circuit. When calculating a Thévenin-equivalent voltage, the voltage divider principle is often useful, by declaring one terminal to be V_{out} and the other terminal to be at the ground point.

The Thévenin-equivalent resistance R_{Th} is the resistance measured across points A and B "looking back" into the circuit. It is important to first replace all voltage- and current-sources with their internal resistances. For an ideal voltage source, this means replace the voltage source with a short circuit. For an ideal current source, this means replace the current source with an open circuit. Resistance can then be calculated across the terminals using the formulae for series and parallel circuits. This method is valid only for circuits with independent sources. If there are dependent sources in the circuit, another method must be used such as connecting a test source across A and B and calculating the voltage across or current through the test source.

Note that the replacement of voltage and current sources do the *opposite* of what the sources themselves are meant to do. A voltage source creates a difference of electric potential between its terminals; its replacement in Thévenin's theorem resistance calculations, a short circuit, equalizes potential. Likewise, a current source's aim is to generate a certain amount of current, whereas an open circuit stops electric flow altogether.

Example

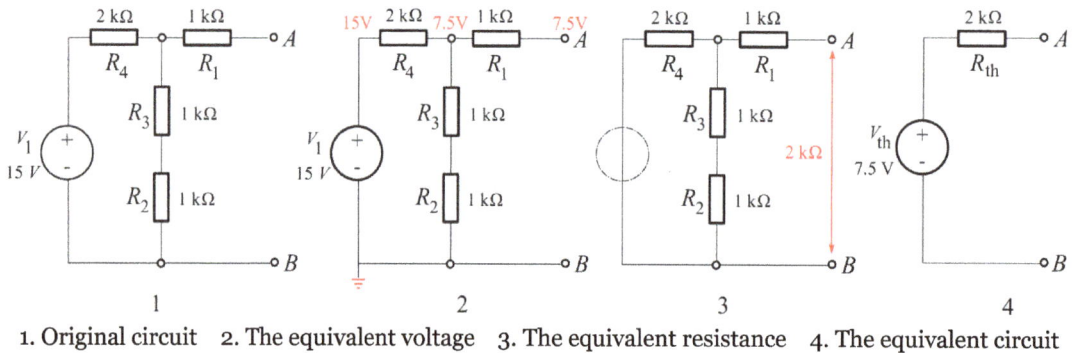

1. Original circuit   2. The equivalent voltage   3. The equivalent resistance   4. The equivalent circuit

In the example, calculating the equivalent voltage:

$$V_{Th} = \frac{R_2 + R_3}{(R_2 + R_3) + R_4} \cdot V_1$$

$$= \frac{1k\Omega + 1k\Omega}{(1k\Omega + 1k\Omega) + 2k\Omega} \cdot 15V$$

$$= \frac{1}{2} \cdot 15V = 7.5V$$

(notice that R_1 is not taken into consideration, as above calculations are done in an open circuit condition between A and B, therefore no current flows through this part, which means there is no current through R_1 and therefore no voltage drop along this part)

Calculating equivalent resistance:

$$R_{Th} = R_1 + \left[(R_2 + R_3) \| R_4 \right]$$

$$= 1k\Omega + \left[(1k\Omega + 1k\Omega) \backslash 2k\Omega \right]$$

$$= 1k\Omega + \left(\frac{1}{(1k\Omega + 1k\Omega)} + \frac{1}{(2k\Omega)} \right) = 2k\Omega.$$

Conversion to a Norton Equivalent

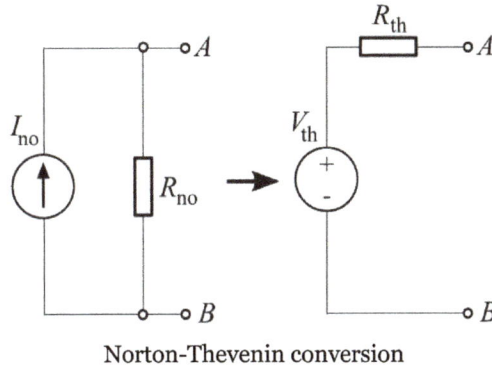

Norton-Thevenin conversion

A Norton equivalent circuit is related to the Thévenin equivalent by

$$R_{\text{Th}} = R_{\text{No}}$$

$$V_{\text{Th}} = I_{\text{No}} R_{\text{No}}$$

$$I_{\text{No}} = V_{\text{Th}} / R_{\text{Th}}.$$

Practical Limitations

- Many circuits are only linear over a certain range of values, thus the Thévenin equivalent is valid only within this linear range.

- The Thévenin equivalent has an equivalent I–V characteristic only from the point of view of the load.

- The power dissipation of the Thévenin equivalent is not necessarily identical to the power dissipation of the real system. However, the power dissipated by an external resistor between the two output terminals is the same regardless of how the internal circuit is implemented.

A proof of the Theorem

The proof involves two steps. First use superposition theorem to construct a solution. Then, use uniqueness theorem to show the solution is unique. The second step is usually implied. First, using the superposition theorem, in general for any linear "black box" circuit which contains voltage sources and resistors, one can always write down its voltage as a linear function of the corresponding current as follows

$$V = V_{\text{Eq}} - Z_{\text{Eq}} I$$

where the first term reflects the linear summation of contributions from each voltage source, while the second term measures the contribution from all the resistors. The above argument is due to the fact that the voltage of the black box for a given current I is identical to the linear superposition of the solutions of the following problems: (1) to leave the black box open circuited but activate individual voltage source one at a time and, (2) to short circuit all the voltage sources but feed the circuit with a certain ideal voltage source so that the resulting current exactly reads I (or an ideal current source of current I). Once the above expression is established, it is straightforward to show that V_{Eq} and Z_{Eq} are the single voltage source and the single series resistor in question.

Constant Phase Element

A constant phase element is an equivalent electrical circuit component that models the behaviour of a double layer, that is an imperfect capacitor.

Constant phase elements are also used in equivalent circuit modelling and data fitting of electrochemical impedance spectroscopy data.

A constant phase element also appears currently in modeling the behaviour of the imperfect dielectrics. The generalization in the fields of imperfect electrical resistances, capacitances and inductances leads to the general "Phasance" concept: http://fr.scribd.com/doc/71923015/The-Phasance-Concept

General Equation

The electrical impedance can be calculated:

$$Z_{CPE} = \frac{1}{Y_{CPE}} = \frac{1}{Q_0 \omega^n} e^{-\frac{\pi}{2}ni}$$

where the CPE admittance is: $Y_{CPE} = Q_0(\omega i)^n$ and Q_0 and n (0<n<1) are frequency independent.

Q_0 = 1/|Z| at ω = 1 rad/s

The constant phase is always –(90*n)°, also with n from 0 to 1. The case n = 1 describes an ideal capacitor while the case n = 0 describes a pure resistor.

Linear Integrated Circuit

An analog chip is a set of miniature electronic analog circuits formed on a single piece of semiconductor material.

Description

The voltage and current at specified points in the circuits of analog chips vary continuously in time. In contrast, digital chips only use and create voltages or currents at discrete levels, with no intermediate values. In addition to Transistors, analog chips often have a larger number of passive elements (Inductor/Capacitors/Resistors) than digital chips typically do. Inductors tend to be avoided because of their large size, and a transistor and capacitor together can do the work of an inductor. (When this method is used in a CFL, you get an electronic ballast.)

Analog chips may also contain digital logic elements to replace some analog functions, or to allow the chip to communicate with a microprocessor. For this reason and since logic is commonly implemented using CMOS technology, these chips use BiCMOS processes by companies such as Freescale, Texas Instruments, STMicroelectronics and others. This is known as mixed signal processing and allows a designer to incorporate more functions in the chip. Some of the benefits include load protection, reduced parts count and higher reliability.

Pure analog chips in information processing have been mostly replaced with digital chips. Analog chips are still required for wideband signals on account of sampling rate requirements, high power applications and at the transducer interfaces. Research and industry in the field continues to grow and prosper. Some examples of long-lived and well-known analog chips are the 741 Operational Amplifier, and the 555 timer.

Power supply chips are also considered to be analog chips. Their main purpose is to produce a well-regulated output voltage supply for other chips in the system. Since all electronic systems require electrical power, power supply ICs PMICs are important elements of those systems.

Important basic building blocks of analog chip design include:

1. current sources

2. current mirrors

3. differential amplifier

4. bandgap references.

All the above circuit building blocks can be implemented using Bipolar technology as well as Metal-Oxide-Silicon(MOS) technology. MOS Band gap references use lateral (poor) bipolar transistors for their functioning.

People who have specialized in this field include Bob Widlar, Bob Pease, Hans R. Camenzind, George Erdi, and Barrie Gilbert among others.

References

- Rao, B. Visvesvara; et al. (2012). Electronic Circuit Analysis. India: Pearson Education India. p. 13.6. ISBN 9332511748.

- Huurdeman, Anton A. (2003). The Worldwide History of Telecommunications. U.S.: Wiley-IEEE. pp. 199–200. ISBN 0-471-20505-2.

4

Various Circuit Theorems

The various circuit theorems discussed in this text are ports, generators, Norton's theorem, Miller's theorem and extra element theorem. The topics discussed in the chapter are of great importance to broaden the existing knowledge on circuit theorems.

Port (Circuit Theory)

In electrical circuit theory, a port is a pair of terminals connecting an electrical network or circuit to an external circuit, a point of entry or exit for electrical energy. A port consists of two nodes (terminals) connected to an outside circuit, that meets the *port condition*; the currents flowing into the two nodes must be equal and opposite.

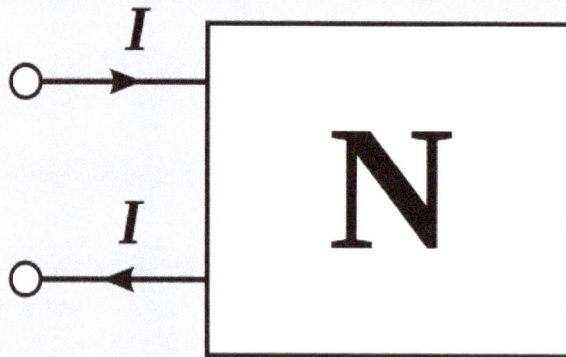

Network N has a port connecting it to an external circuit. The port meets the port condition because the current I entering one terminal of the port is equal to the current exiting the other.

The use of ports helps to reduce the complexity of circuit analysis. Many common electronic devices and circuit blocks, such as transistors, transformers, electronic filters, and amplifiers, are analyzed in terms of ports. In multiport network analysis, the circuit is regarded as a "black box" connected to the outside world through its ports. The ports are points where input signals are applied or output signals taken. Its behavior is completely specified by a matrix of parameters relating the voltage and current at its ports, so the internal makeup or design of the circuit need not be considered, or even known, in determining the circuit's response to applied signals.

The concept of ports can be extended to waveguides, but the definition in terms of current is not appropriate and the possible existence of multiple waveguide modes must be accounted for.

Port Condition

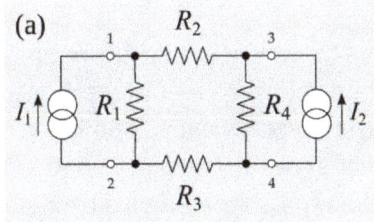

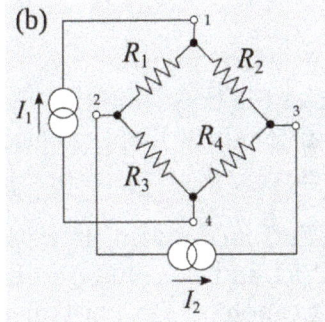

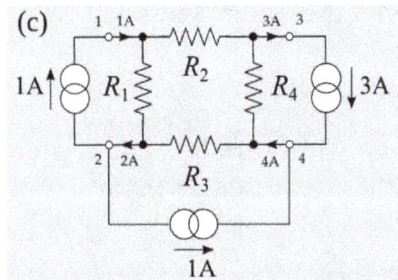

Simple resistive network with three possible port arrangements: (a) Pole pairs (1, 2) and (3, 4) are ports; (b) pole pairs (1, 4) and (2, 3) are ports; (c) no pair of poles are ports

Any node of a circuit that is available for connection to an external circuit is called a pole (or terminal if it is a physical object). The port condition is that a pair of poles of a circuit is considered a port if and only if the current flowing into one pole from outside the circuit is equal to the current flowing out of the other pole into the external circuit. Equivalently, the algebraic sum of the currents flowing into the two poles from the external circuit must be zero.

It cannot be determined if a pair of nodes meets the port condition by analysing the internal properties of the circuit itself. The port condition is dependent entirely on the external connections of the circuit. What are ports under one set of external circumstances may well not be ports under another. Consider the circuit of four resistors in the figure for example. If generators are connected to the pole pairs (1, 2) and (3, 4) then those two pairs are ports and the circuit is a box attenuator. On the other hand, if generators are connected to pole pairs (1, 4) and (2, 3) then those pairs are ports, the pairs (1, 2) and (3, 4) are no longer ports, and the circuit is a bridge circuit.

It is even possible to arrange the inputs so that *no* pair of poles meets the port condition. However, it is possible to deal with such a circuit by splitting one or more poles into a number of separate poles joined to the same node. If only one external generator terminal is connected to each pole (whether a split pole or otherwise) then the circuit can again be analysed in terms of ports. The most common arrangement of this type is to designate one pole of an n-pole circuit as the common and split it into $n-1$ poles. This latter form is especially useful for unbalanced circuit topologies and the resulting circuit has $n-1$ ports.

In the most general case, it is possible to have a generator connected to every pair of poles, that is, ${}^nC_2$ generators, then every pole must be split into $n-1$ poles. For instance, in the figure example (c), if the poles 2 and 4 are each split into two poles each then the circuit can be described as a 3-port. However, it is also possible to connect generators to pole pairs (1, 3), (1, 4), and (3, 2) making ${}^4C_2$ = 6 generators in all and the circuit has to be treated as a 6-port.

One-ports

Any two-pole circuit is guaranteed to meet the port condition by virtue of Kirchhoff's current law and they are therefore one-ports unconditionally. All of the basic electrical elements (inductance, resistance, capacitance, voltage source, current source) are one-ports, as is a general impedance.

Study of one-ports is an important part of the foundation of network synthesis, most especially in filter design. Two-element one-ports (that is RC, RL and LC circuits) are easier to synthesise than the general case. For a two-element one-port Foster's canonical form or Cauer's canonical form can be used. In particular, LC circuits are studied since these are lossless and are commonly used in filter design.

Two-ports

Linear two port networks have been widely studied and a large number of ways of representing them have been developed. One of these representations is the z-parameters which can be described in matrix form by;

$$\begin{bmatrix} V_1 \\ V_2 \end{bmatrix} = \begin{bmatrix} z_{11} & z_{12} \\ z_{21} & z_{22} \end{bmatrix} \begin{bmatrix} I_1 \\ I_2 \end{bmatrix}$$

where V_n and I_n are the voltages and currents respectively at port n. Most of the other descriptions of two-ports can likewise be described with a similar matrix but with a different arrangement of the voltage and current column vectors.

Common circuit blocks which are two-ports include amplifiers, attenuators and filters.

Multiports

Coaxial circulators. Circulators have at least three ports

In general, a circuit can consist of any number of ports—a multiport. Some, but not all, of the two-port parameter representations can be extended to arbitrary multiports. Of the voltage and current based matrices, the ones that can be extended are z-parameters and y-parameters. Neither of these are suitable for use at microwave frequencies because voltages and currents are not convenient to measure in formats using conductors and are not relevant at all in waveguide formats. Instead, s-parameters are used at these frequencies and these too can be extended to an arbitrary number of ports.

Circuit blocks which have more than two ports include directional couplers, power splitters, circulators, diplexers, duplexers, multiplexers, hybrids and directional filters.

RF and Microwave

RF and microwave circuit topologies are commonly unbalanced circuit topologies such as coaxial or microstrip. In these formats, one pole of each port in a circuit is connected to a common node such as a ground plane. It is assumed in the circuit analysis that all these commoned poles are at the same potential and that current is sourced to or sunk into the ground plane that is equal and opposite to that going into the other pole of any port. In this topology a port is treated as being just a single pole. The corresponding balancing pole is imagined to be incorporated into the ground plane.

The one-pole representation of a port will start to fail if there are significant ground plane loop currents. The assumption in the model is that the ground plane is perfectly conducting and that there is no potential difference between two locations on the ground plane. In reality, the ground plane is not perfectly conducting and loop currents in it will cause potential differences. If there is a potential difference between the commoned poles of two ports then the port condition is broken and the model is invalid.

Waveguide

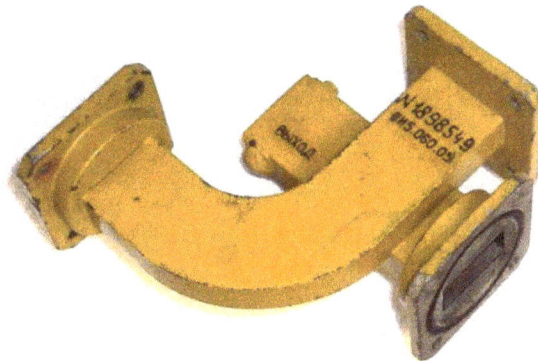

A Moreno coupler, a type of waveguide directional coupler. Directional couplers have four ports. This one has one port permanently terminated internally with a matched load, so only three ports are visible. The ports are the openings in the centres of the waveguide flanges

The idea of ports can be (and is) extended to waveguide devices, but a port can no longer be defined in terms of circuit poles because in waveguides the electromagnetic waves are not guided by electric conductors. They are, instead, guided by the walls of the waveguide. Thus, the concept of a circuit conductor pole does not exist in this format. Ports in waveguides consist of an aperture or break in the waveguide through which the electromagnetic waves can pass. The bounded plane through which the wave passes is the definition of the port.

Waveguides have an additional complication in port analysis in that it is possible (and sometimes desirable) for more than one waveguide mode to exist at the same time. In such cases, for each physical port, a separate port must be added to the analysis model for each of the modes present at that physical port.

Other Energy Domains

The concept of ports can be extended into other energy domains. The generalised definition of a port is a place where energy can flow from one element or subsystem to another element or subsystem. This generalised view of the port concept helps to explain why the port condition is so defined in electrical analysis. If the algebraic sum of the currents is not zero, such as in example diagram (c), then the energy delivered from an external generator is not equal to the energy entering the pair of circuit poles. The energy transfer at that place is thus more complex than a simple flow from one subsystem to another and does not meet the generalised definition of a port.

The port concept is particularly useful where multiple energy domains are involved in the same system and a unified, coherent analysis is required such as with mechanical-electrical analogies or bond graph analysis. Connection between energy domains is by means of transducers. A transducer may be a one-port as viewed by the electrical domain, but with the more generalised definition of *port* it is a two-port. For instance, a mechanical actuator has one port in the electrical domain and one port in the mechanical domain. Transducers can be analysed as two-port networks in the same way as electrical two-ports. However, the variables at the two ports will be different and the two-port parameters will be a mixture of two energy domains. For instance, in the actuator example, the z-parameters will include one electrical impedance, one mechanical impedance, and two transimpedances that are ratios of one electrical and one mechanical variable.

Generator (Circuit Theory)

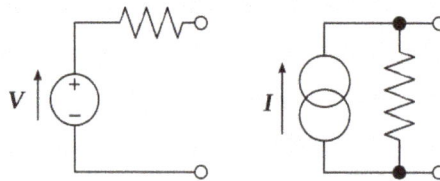

Non-ideal voltage source model (left) and non-ideal current source model (right)

A generator in electrical circuit theory is one of two ideal elements: an ideal voltage source, or an ideal current source. These are two of the fundamental elements in circuit theory. Real electrical generators are most commonly modelled as a non-ideal source consisting of a combination of an ideal source and a resistor. Voltage generators are modelled as an ideal voltage source in series with a resistor. Current generators are modelled as an ideal current source in parallel with a resistor. The resistor is referred to as the internal resistance of the source. Real world equipment may not perfectly follow these models, especially at extremes of loading (both high and low) but for most purposes they suffice.

The two models of non-ideal generators are interchangeable, either can be used for any given generator. Thévenin's theorem allows a non-ideal current source model to be converted to a non-ideal voltage source model and Norton's theorem allows a non-ideal voltage source model to be converted to a non-ideal current source model. Both models are equally valid, but the voltage source model is more applicable to when the internal resistance is low (that is, much lower than the load impedance) and the current source model is more applicable when the internal resistance is high (compared to the load).

Symbols

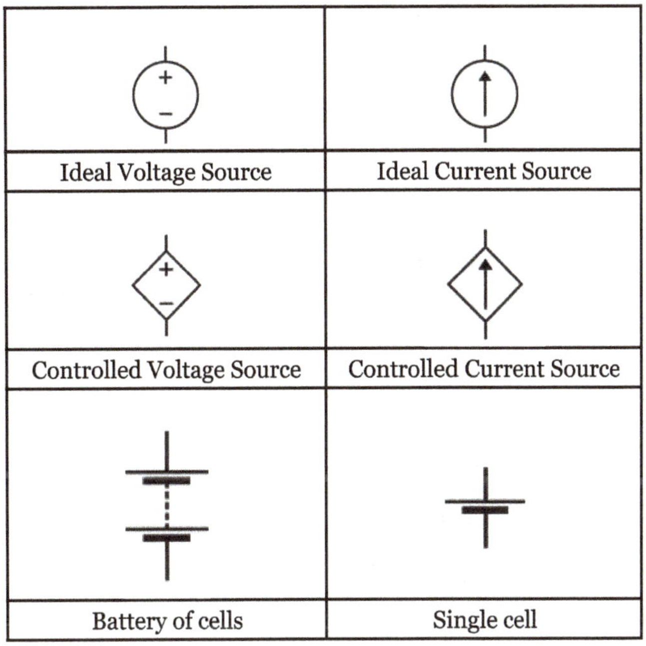

Ideal Voltage Source	Ideal Current Source
Controlled Voltage Source	Controlled Current Source
Battery of cells	Single cell

Symbols used for ideal sources

Symbols commonly used for ideal sources are shown in the figure. Symbols do vary from region to region and time period to time period. Another common symbol for a current source is two interlocking circles.

Dependent Sources

A dependent source is one in which the voltage or current of the source output is dependent on another voltage or current elsewhere in the circuit. There are thus four possible types: current dependent voltage source, voltage dependent voltage source, current dependent current source and voltage dependent current source. Non-ideal dependent sources can be modelled with the addition of an impedance in the same way as non-dependent sources. These elements are widely used to model the function of two-port networks; one generator is needed for each port and it is dependent on either voltage or current at the other port. The models are an example of black box modelling, that is, they are quite unrelated to what is physically inside the device but correctly model the device's function. There are a number of these two-port models, differing only in the type of generator required to represent them. This kind of model is particularly useful for modelling the behaviour of transistors.

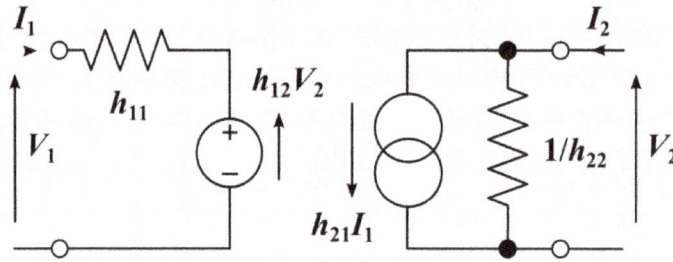

Dependent sources used to represent a two-port network in h-parameters

The model used to represent h-parameters is shown in the figure. h-parameters are frequently used in transistor data sheets to specify the device. The h-parameters are defined as the matrix

$$\begin{bmatrix} V_1 \\ I_2 \end{bmatrix} = \begin{bmatrix} h_{11} & h_{12} \\ h_{21} & h_{22} \end{bmatrix} \begin{bmatrix} I_1 \\ V_2 \end{bmatrix}$$

where the voltage and current variables are as shown in the figure. The circuit model using dependent generators is just an alternative way of representing this matrix.

Norton's Theorem

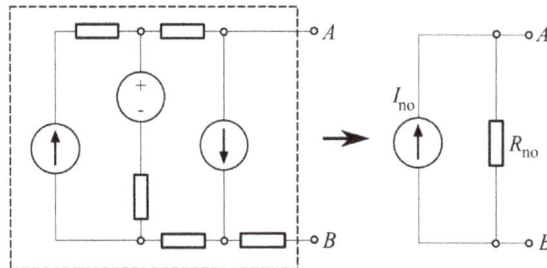

Any black box containing resistances only and voltage and current sources can be replaced by an equivalent circuit consisting of an equivalent current source in parallel connection with an equivalent resistance.

Edward Lawry Norton

Known in Europe as the Mayer–Norton theorem, Norton's theorem holds, to illustrate in DC circuit theory terms, that:

- Any linear electrical network with voltage and current sources and only resistances can be replaced at terminals A-B by an equivalent current source I_{no} in parallel connection with an equivalent resistance R_{no}.

- This equivalent current I_{no} is the current obtained at terminals A-B of the network with terminals A-B short circuited.

- This equivalent resistance R_{no} is the resistance obtained at terminals A-B of the network with all its voltage sources short circuited and all its current sources open circuited.

For AC systems the theorem can be applied to reactive impedances as well as resistances.

The Norton equivalent circuit is used to represent any network of linear sources and impedances at a given frequency.

Norton's theorem and its dual, Thévenin's theorem, are widely used for circuit analysis simplification and to study circuit's initial-condition and steady-state response.

Norton's theorem was independently derived in 1926 by Siemens & Halske researcher Hans Ferdinand Mayer (1895–1980) and Bell Labs engineer Edward Lawry Norton (1898–1983).

To find the equivalent,

1. Find the Norton current I_{no}. Calculate the output current, I_{AB}, with a short circuit as the load (meaning 0 resistance between A and B). This is I_{no}.

2. Find the Norton resistance R_{no}. When there are no dependent sources (all current and voltage sources are independent), there are two methods of determining the Norton impedance R_{no}.

    - Calculate the output voltage, V_{AB}, when in open circuit condition (i.e., no load resistor – meaning infinite load resistance). R_{no} equals this V_{AB} divided by I_{no}.

    or

    - Replace independent voltage sources with short circuits and independent current sources with open circuits. The total resistance across the output port is the Norton impedance R_{no}.

This is equivalent to calculating the Thevenin resistance.

   However, when there are dependent sources, the more general method must be used. This method is not shown below in the diagrams.

    - Connect a constant current source at the output terminals of the circuit with a value of 1 Ampere and calculate the voltage at its terminals. This voltage divided by the 1 A current is the Norton impedance R_{no}. This method must be used if the circuit contains dependent sources, but it can be used in all cases even when there are no dependent sources.

Example of a Norton Equivalent Circuit

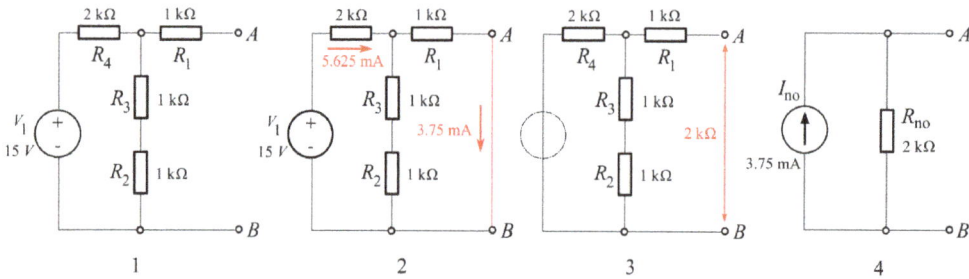

1. The original circuit   2. Calculating the equivalent output current   3. Calculating the equivalent resistance   4. Design the equivalent circuit

In the example, the total current I_{total} is given by:

$$I_{\text{total}} = \frac{15V}{2k\Omega + 1k\Omega\backslash(1k\Omega + 1k\Omega)} = 5.625mA.$$

The current through the load is then, using the current divider rule:

$$I_{no} = \frac{1k\Omega + 1k\Omega}{(1k\Omega + 1k\Omega + 1k\Omega)} \cdot I_{\text{total}}$$

$$= 2/3 \cdot 5.625mA = 3.75mA.$$

And the equivalent resistance looking back into the circuit is:

$$R_{no} = 1k\Omega + (2k\Omega\backslash(1k\Omega + 1k\Omega)) = 2k\Omega.$$

So the equivalent circuit is a 3.75 mA current source in parallel with a 2 kΩ resistor.

Conversion to a Thévenin Equivalent

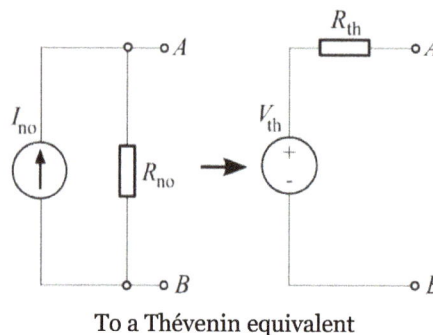

To a Thévenin equivalent

A Norton equivalent circuit is related to the Thévenin equivalent by the equations:

$$R_{th} = R_{no}$$

$$V_{th} = I_{no}R_{no}$$

$$\frac{V_{th}}{R_{th}} = I_{no}$$

Queueing Theory

The passive circuit equivalent of "Norton's theorem" in queuing theory is called the Chandy Herzog Woo theorem. In a reversible queueing system, it is often possible to replace an uninteresting subset of queues by a single (FCFS or PS) queue with an appropriately chosen service rate.

Miller Theorem

The Miller theorem refers to the process of creating equivalent circuits. It asserts that a floating impedance element, supplied by two voltage sources connected in series, may be split into two grounded elements with corresponding impedances. There is also a dual Miller theorem with regards to impedance supplied by two current sources connected in parallel. The two versions are based on the two Kirchhoff's circuit laws.

Miller theorems are not only pure mathematical expressions. These arrangements explain important circuit phenomena about modifying impedance (Miller effect, virtual ground, bootstrapping, negative impedance, etc.) and help in designing and understanding various commonplace circuits (feedback amplifiers, resistive and time-dependent converters, negative impedance converters, etc.). The theorems are useful in 'circuit analysis' especially for analyzing circuits with feedback and certain transistor amplifiers at high frequencies.

There is a close relationship between Miller theorem and Miller effect: the theorem may be considered as a generalization of the effect and the effect may be thought as of a special case of the theorem.

Miller Theorem (For Voltages)

Definition

The Miller theorem establishes that in a linear circuit, if there exists a branch with impedance Z, connecting two nodes with nodal voltages V_1 and V_2, we can replace this branch by two branches connecting the corresponding nodes to ground by impedances respectively $Z/(1 - K)$ and $KZ/(K - 1)$, where $K = V_2/V_1$. The Miller theorem may be proved by using the equivalent two-port network technique to replace the two-port to its equivalent and by applying the source absorption theorem. This version of the Miller theorem is based on Kirchhoff's voltage law; for that reason, it is named also *Miller theorem for voltages*.

Explanation

Schematic missing

Miller theorem implies that an impedance element is supplied by two arbitrary (not necessarily dependent) voltage sources that are connected in series through the common ground. In practice, one of them acts as a main (independent) voltage source with voltage V_1 and the other – as an additional (linearly dependent) voltage source with voltage $V_2 = KV_1$. The idea of Miller theorem (modifying circuit impedances seen from the sides of the input and output sources) is revealed below by comparing the two situations – without and with connecting an additional voltage source V_2.

If V_2 was zero (there was not a second voltage source or the right end of the element with impedance Z was just grounded), the input current flowing through the element would be determined, according to Ohm's law, only by V_1

$$I_{in0} = \frac{V_1}{Z}$$

and the input impedance of the circuit would be

$$I_{in0} = \frac{V_1}{Z}$$

As a second voltage source is included, the input current depends on both the voltages. According to its polarity, V_2 is subtracted from or added to V_1; so, the input current decreases/increases

$$I_{in} = \frac{V_1 - V_2}{Z} = \frac{(1-K)}{Z} V_1 = (1-K)I_{in0}$$

and the input impedance of the circuit seen from the side of the input source accordingly increases/decreases

$$Z_{in} = \frac{V_1}{I_{in}} = \frac{Z}{1-K}$$

So, Miller theorem expresses the fact that *connecting a second voltage source with proportional voltage $V_2 = KV_1$ in series with the input voltage source changes the effective voltage, the current and respectively, the circuit impedance seen from the side of the input source*. Depending on the polarity, V_2 acts as a supplemental voltage source helping or opposing the main voltage source to pass the current through the impedance.

Besides by presenting the combination of the two voltage sources as a new composed voltage source, the theorem may be explained by *combining the actual element and the second voltage source into a new virtual element with dynamically modified impedance*. From this viewpoint, V_2 is an additional voltage that artificially increases/decreases the voltage drop V_z across the impedance Z thus decreasing/increasing the current. The proportion between the voltages determines the value of the obtained impedance (see the tables below) and gives in total six groups of typical applications.

Subtracting V_2 from V_1				
V_2 vs V_1	$V_2 = 0$	$0 < V_2 < V_1$	$V_2 = V_1$	$V_2 > V_1$
Impedance	normal	increased	infinite	negative with current inversion

Adding V_2 to V_1				
V_2 vs V_z	$V_2 = 0$	$0 < V_2 < V_z$	$V_2 = V_z$	$V_2 > V_z$
Impedance	normal	decreased	zero	negative with voltage inversion

The circuit impedance, seen from the side of the output source, may be defined similarly, if the voltages V_1 and V_2 are swapped and the coefficient K is replaced by $1/K$

$$Z_{in2} = \frac{KZ}{K-1}.$$

Implementation

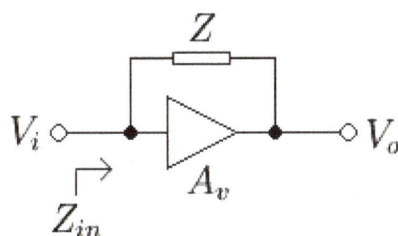

A typical implementation of Miller theorem based on a single-ended voltage amplifier

Most frequently, the Miller theorem may be observed in, and implemented by, an arrangement consisting of an element with impedance Z connected between the two terminals of a grounded general linear network. Usually, a voltage amplifier with gain of $A_V = K$ serves as such a linear network, but also other devices can play this role: a man and a potentiometer in a potentiometric null-balance meter, an electromechanical integrator (servomechanisms using potentiometric feedback sensors), etc.

In the amplifier implementation, the input voltage V_i serves as V_1 and the output voltage V_o – as V_2. In many cases, the input voltage source has some internal impedance Z_{int} or an additional input impedance is connected that, in combination with Z, introduces a feedback. Depending on the kind of amplifier (non-inverting, inverting or differential), the feedback can be positive, negative or mixed.

The Miller amplifier arrangement has two aspects:

- the amplifier may be thought as an additional voltage source converting the actual impedance into a *virtual impedance* (the amplifier modifies the impedance of the actual element)

- the virtual impedance may be thought as an element connected in parallel to the amplifier input (the virtual impedance modifies the amplifier input impedance).

Applications

The introduction of an impedance that connects amplifier input and output ports adds a great deal of complexity in the analysis process. Miller theorem helps reduce the complexity in some circuits particularly with feedback by converting them to simpler equivalent circuits. But Miller theorem is not only an effective tool for creating equivalent circuits; it is also a powerful tool for designing and understanding circuits based on *modifying impedance by additional voltage*. Depending on the polarity of the output voltage versus the input voltage and the proportion between their magnitudes, there are six groups of typical situations. In some of them, the Miller phenomenon appears as desired (bootstrapping) or undesired (Miller effect) unintentional effects; in other cases it is intentionally introduced.

Applications Based on Subtracting V_2 from V_1

In these applications, the output voltage V_o is inserted with an opposite polarity in respect to the input voltage V_i travelling along the loop (but in respect to ground, the polarities are the same). As a result, the effective voltage across, and the current through, the impedance decrease; the input impedance increases.

Increased impedance is implemented by a non-inverting amplifier with gain of $0 < A_v < 1$. The (magnitude of) output voltage is less than the input voltage V_i and partially neutralizes it. Examples are imperfect voltage followers (emitter, source, cathode follower, etc.) and amplifiers with series negative feedback (emitter degeneration), whose input impedance is moderately increased.

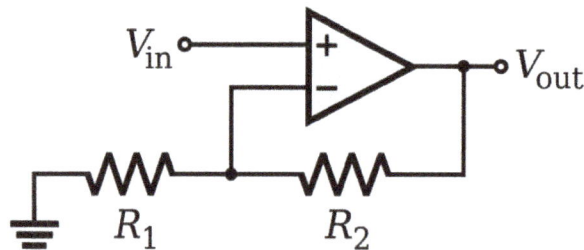

The op-amp non-inverting amplifier is a typical circuit with series negative feedback based on the Miller theorem, where the op-amp differential input impedance is apparently increased up to infinity

Infinite impedance uses a non-inverting amplifier with $A_v = 1$. The output voltage is equal to the input voltage V_i and completely neutralizes it. Examples are potentiometric null-balance meters and op-amp followers and amplifiers with series negative feedback (op-amp follower and non-inverting amplifier) where the circuit input impedance is enormously increased. This technique is referred to as bootstrapping and is intentionally used in biasing circuits, input guarding circuits, etc.

Negative impedance obtained by current inversion is implemented by a non-inverting amplifier with $A_v > 1$. The current changes its direction, as the output voltage is higher than the input voltage. If the input voltage source has some internal impedance Z_{int} or if it is connected through another impedance element, a positive feedback appears. A typical application is the negative impedance converter with current inversion (INIC) that uses both negative and positive feedback (the negative feedback is used to realize a non-inverting amplifier and the positive feedback – to modify the impedance).

Applications Based on Adding V_2 to V_1

In these applications, the output voltage V_o is inserted with the same polarity in respect to the input voltage V_i travelling along the loop (but in respect to ground, the polarities are opposite).

As a result, the effective voltage across and the current through the impedance increase; the input impedance decreases.

Decreased impedance is implemented by an inverting amplifier having some moderate gain, usually $10 < A_v < 1000$. It may be observed as an undesired Miller effect in common-emitter, common-source and *common-cathode* amplifying stages where effective input capacitance is increased. Frequency compensation for general purpose operational amplifiers and transistor Miller integrator are examples of useful usage of the Miller effect.

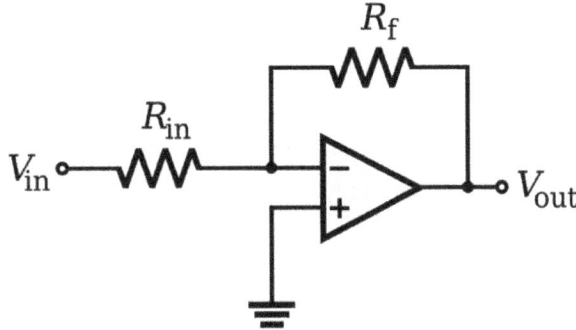

The op-amp inverting amplifier is a typical circuit, with parallel negative feedback, based on the Miller theorem, where the op-amp differential input impedance is apparently decreased up to zero

Zeroed impedance uses an inverting (usually op-amp) amplifier with enormously high gain $A_v \rightarrow \infty$. The output voltage is almost equal to the voltage drop V_z across the impedance and completely neutralizes it. The circuit behaves as a short connection and a virtual ground appears at the input; so, it should not be driven by a constant voltage source. For this purpose, some circuits are driven by a constant current source or by a real voltage source with internal impedance: current-to-voltage converter (transimpedance amplifier), capacitive integrator (named also current integrator or charge amplifier), resistance-to-voltage converter (a resistive sensor connected in the place of the impedance Z).

The rest of them have additional impedance connected in series to the input: voltage-to-current converter (transconductance amplifier), inverting amplifier, summing amplifier, inductive integrator, capacitive differentiator, resistive-capacitive integrator, capacitive-resistive differentiator, inductive-resistive differentiator, etc. The inverting integrators from this list are examples of useful and desired applications of the Miller effect in its extreme manifestation.

In all these *op-amp inverting circuits with parallel negative feedback*, the input current is increased to its maximum. It is determined only by the input voltage and the input impedance according to Ohm's law; it does not depend on the impedance Z.

Negative impedance with voltage inversion is implemented by applying both negative and positive feedback to an op-amp amplifier with a differential input. The input voltage source has to have internal impedance $Z_{int} > 0$ or it has to be connected through another impedance element to the input. Under these conditions, the input voltage V_i of the circuit changes its polarity as the output voltage exceeds the voltage drop V_z across the impedance ($V_i = V_z - V_o < 0$).

A typical application is a negative impedance converter with voltage inversion (VNIC). It is interesting that the circuit input voltage has the same polarity as the output voltage, although it is applied to the inverting op-amp input; the input source has an opposite polarity to both the circuit input and output voltages.

Generalization of Miller Arrangement

The original Miller effect is implemented by capacitive impedance connected between the two nodes. Miller theorem generalizes Miller effect as it implies arbitrary impedance Z connected between the nodes. It is supposed also a constant coefficient K; then the expressions above are valid. But modifying properties of Miller theorem exist even when these requirements are violated and this arrangement can be generalized further by dynamizing the impedance and the coefficient.

Non-linear element. Besides impedance, Miller arrangement can modify the IV characteristic of an arbitrary element. The circuit of a diode log converter is an example of a non-linear virtually zeroed resistance where the logarithmic forward IV curve of a diode is transformed to a vertical straight line overlapping the Y axis.

Not constant coefficient. If the coefficient K varies, some exotic virtual elements can be obtained. A gyrator circuit is an example of such a virtual element where the resistance R_L is modified so that to mimic inductance, capacitance or inversed resistance.

Dual Miller Theorem (for Currents)

Definition

There is also a dual version of Miller theorem that is based on Kirchhoff's current law (*Miller theorem for currents*): if there is a branch in a circuit with impedance Z connecting a node, where two currents I_1 and I_2 converge to ground, we can replace this branch by two conducting the referred currents, with impedances respectively equal to $(1 + \alpha)Z$ and $(1 + \alpha)Z/\alpha$, where $\alpha = I_2/I_1$. The dual theorem may be proved by replacing the two-port network by its equivalent and by applying the source absorption theorem.

Explanation

Dual Miller theorem actually expresses the fact that *connecting a second current source producing proportional current $I_2 = KI_1$ in parallel with the main input source and the impedance element changes the current flowing through it, the voltage and accordingly, the circuit impedance seen from the side of the input source.* Depending on the direction, I_2 acts as a supplemental current source helping or opposing the main current source I_1 to create voltage across the impedance. The combination of the actual element and the second current source may be thought as of a new virtual element with dynamically modified impedance.

Implementation

Dual Miller theorem is usually implemented by an arrangement consisting of two voltage sources supplying the grounded impedance Z through floating impedances. The combinations of the voltage sources and belonging impedances form the two current sources – the main and the auxiliary one. As in the case of the main Miller theorem, the second voltage is usually produced by a voltage amplifier. Depending on the kind of the amplifier (inverting, non-inverting or differential) and the gain, the circuit input impedance may be virtually increased, infinite, decreased, zero or negative.

Applications

As the main Miller theorem, besides helping circuit analysis process, the dual version is a powerful tool for designing and understanding circuits based on modifying impedance by additional current. Typical applications are some exotic circuits with negative impedance as load cancellers, capacitance neutralizers, Howland current source and its derivative Deboo integrator. In the last example, the Howland current source consists of an input voltage source V_{IN}, a positive resistor R, a load (the capacitor C acting as impedance Z) and a negative impedance converter INIC ($R_1 = R_2 = R_3 = R$ and the op-amp). The input voltage source and the resistor R constitute an imperfect current source passing current I_R through the load. The INIC acts as a second current source passing "helping" current I_{-R} through the load. As a result, the total current flowing through the load is constant and the circuit impedance seen by the input source is increased. As a comparison, in a load canceller, the INIC passes all the required current through the load; the circuit impedance seen from the side of the input source (the load impedance) is almost infinite.

List of Specific Applications Based on Miller Theorems

Below is a list of circuit solutions, phenomena and techniques based on the two Miller theorems.

Circuit solutions

- Potentiometric null-balance meter

- Electromechanical data recorders with a potentiometric servo system

- Emitter (source, cathode) follower

- Transistor amplifier with emitter (source, cathode) degeneration

- Transistor bootstrapped biasing circuits

- Transistor integrator

- Common-emitter (common-source, common-cathode) amplifying stages with stray capacitances

- Op-amp follower

- Op-amp non-inverting amplifier

- Op-amp bootstrapped AC follower with high input impedance

- Bilateral current source

- Negative impedance converter with current inversion (INIC)

- Negative impedance load canceller

- Negative impedance input capacitance canceller
- Howland current source
- Deboo integrator
- Op-amp inverting ammeter
- Op-amp voltage-to-current converter (transconductance amplifier)
- Op-amp current-to-voltage converter (transimpedance amplifier)
- Op-amp resistance-to-current converter
- Op-amp resistance-to-voltage converter
- Op-amp inverting amplifier
- Op-amp inverting summer
- Op-amp inverting capacitive integrator (current integrator, charge amplifier)
- Op-amp inverting resistive-capacitive integrator
- Op-amp inverting capacitive differentiator
- Op-amp inverting capacitive-resistive differentiator
- Op-amp inverting inductive integrator
- Op-amp inverting inductive-resistive differentiator, etc.
- Op-amp diode log converter
- Op-amp diode anti-log converter
- Op-amp inverting diode limiter (precision diode)
- Negative impedance converter with voltage inversion (VNIC), etc.

Circuit phenomena and techniques

- Bootstrapping
- Input guarding of high impedance op-amp circuits
- Input-capacitance neutralization
- Virtual ground
- Miller effect
- Frequency op-amp compensation
- Negative impedance
- Load cancelling

Foster's Reactance Theorem

Foster's reactance theorem is an important theorem in the fields of electrical network analysis and synthesis. The theorem states that the reactance of a passive, lossless two-terminal (one-port) network always strictly monotonically increases with frequency. It is easily seen that the reactances of inductors and capacitors individually increase with frequency and from that basis a proof for passive lossless networks generally can be constructed. The proof of the theorem was presented by Ronald Martin Foster in 1924, although the principle had been published earlier by Foster's colleagues at American Telephone & Telegraph.

The theorem can be extended to admittances and the encompassing concept of immittances. A consequence of Foster's theorem is that poles and zeroes of the reactance must alternate with frequency. Foster used this property to develop two canonical forms for realising these networks. Foster's work was an important starting point for the development of network synthesis.

It is possible to construct non-Foster networks using active components such as amplifiers. These can generate an impedance equivalent to a negative inductance or capacitance. The negative impedance converter is an example of such a circuit.

Explanation

Reactance is the imaginary part of the complex electrical impedance. Both capacitors and inductors possess reactance (but of opposite sign) and are frequency dependent. The specification that the network must be passive and lossless implies that there are no resistors (lossless), or amplifiers or energy sources (passive) in the network. The network consequently must consist entirely of inductors and capacitors and the impedance will be purely an imaginary number with zero real part. Foster's theorem applies equally to the admittance of a network, that is the susceptance (imaginary part of admittance) of a passive, lossless one-port monotonically increases with frequency. This result may seem counterintuitive since admittance is the reciprocal of impedance, but is easily proved. If the impedance is

$$Z = iX$$

where X is reactance and i is the imaginary unit, then the admittance is given by

$$Y = \frac{1}{iX} = -i\frac{1}{X} = iB$$

where B is susceptance.

If X is monotonically increasing with frequency then $1/X$ must be monotonically decreasing. $-1/X$ must consequently be monotonically increasing and hence it is proved that B is increasing also.

It is often the case in network theory that a principle or procedure applies equally well to impedance or admittance—reflecting the principle of duality for electric networks. It is convenient in these circumstances to use the concept of immittance, which can mean either impedance or admittance.

The mathematics is carried out without specifying units until it is desired to calculate a specific example. Foster's theorem can thus be stated in a more general form as,

> Foster's theorem (immittance form)
>
> *The imaginary immittance of a passive, lossless one-port strictly monotonically increases with frequency.*

Foster's theorem is quite general. In particular, it applies to distributed element networks, although Foster formulated it in terms of discrete inductors and capacitors. It is therefore applicable at microwave frequencies just as much as it is at lower frequencies.

Examples

Plot of the reactance of an inductor against frequency	Plot of the reactance of a capacitor against frequency
Plot of the reactance of a series *LC* circuit against frequency	Plot of the reactance of a parallel *LC* circuit against frequency

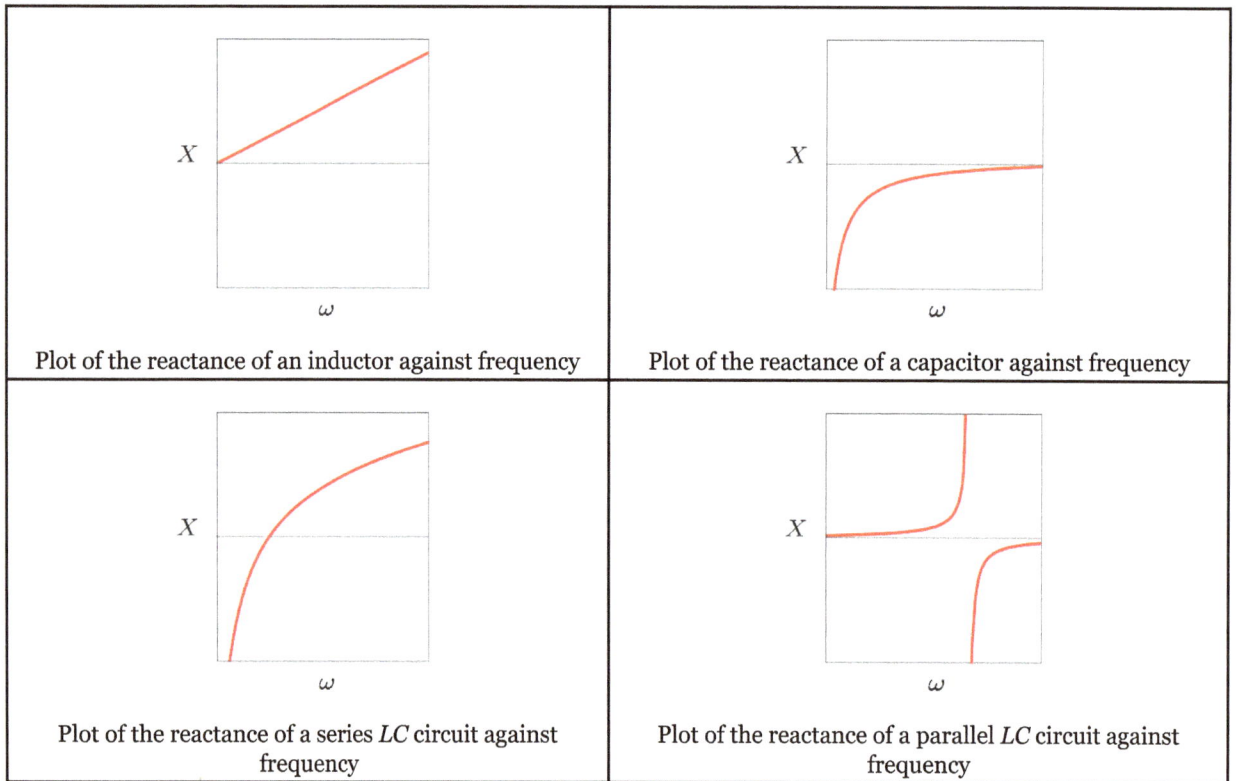

The following examples illustrate this theorem in a number of simple circuits.

Inductor

The impedance of an inductor is given by,

$$Z = i\omega L$$

L is inductance

ω is angular frequency

so the reactance is,

$$X = \omega L$$

which by inspection can be seen to be monotonically (and linearly) increasing with frequency.

Capacitor

The impedance of a capacitor is given by,

$$Z = \frac{1}{i\omega C}$$

　　　　C is capacitance

so the reactance is,

$$X = -\frac{1}{\omega C}$$

which again is monotonically increasing with frequency. The impedance function of the capacitor is identical to the admittance function of the inductor and vice versa. It is a general result that the dual of any immittance function that obeys Foster's theorem will also follow Foster's theorem.

Series Resonant Circuit

A series LC circuit has an impedance that is the sum of the impedances of an inductor and capacitor,

$$Z = i\omega L + \frac{1}{i\omega C} = i\left(\omega L - \frac{1}{\omega C} \right)$$

At low frequencies the reactance is dominated by the capacitor and so is large and negative. This monotonically increases towards zero (the magnitude of the capacitor reactance is becoming smaller). The reactance passes through zero at the point where the magnitudes of the capacitor and inductor reactances are equal (the resonant frequency) and then continues to monotonically increase as the inductor reactance becomes progressively dominant.

Parallel Resonant Circuit

A parallel LC circuit is the dual of the series circuit and hence its admittance function is the same form as the impedance function of the series circuit,

$$Y = i\omega C + \frac{1}{i\omega L}$$

The impedance function is,

$$Z = i\left(\frac{\omega L}{1 - \omega^2 LC} \right)$$

At low frequencies the reactance is dominated by the inductor and is small and positive. This monotonically increases towards a pole at the anti-resonant frequency where the susceptance of the inductor and capacitor are equal and opposite and cancel. Past the pole the reactance is large and negative and increasing towards zero where it is dominated by the capacitance.

Poles and Zeroes

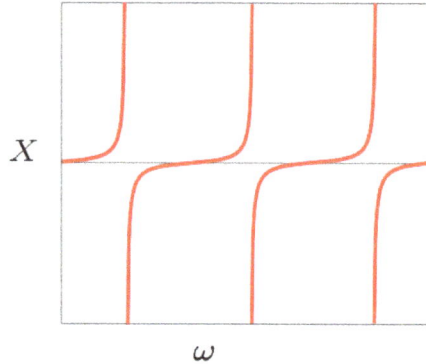

Plot of the reactance of Foster's first form of canonical driving point impedance showing the pattern of alternating poles and zeroes. Three anti-resonators are required to realise this impedance function.

A consequence of Foster's theorem is that the poles and zeroes of any passive immittance function must alternate as frequency increases. After passing through a pole the function will be negative and is obliged to pass through zero before reaching the next pole if it is to be monotonically increasing.

The poles and zeroes of an immittance function completely determine the frequency characteristics of a Foster network. Two Foster networks that have identical poles and zeroes will be equivalent circuits in the sense that their immittance functions will be identical. There can be a scaling factor difference between them (all elements of the immittance multiplied by the same scaling factor) but the *shape* of the two immittance functions will be identical.

Another consequence of Foster's theorem is that the phase of an immittance must monatonically increase with frequency. Consequently, the plot of a Foster immittance function on a Smith chart must always travel around the chart in a clockwise direction with increasing frequency.

Realisation

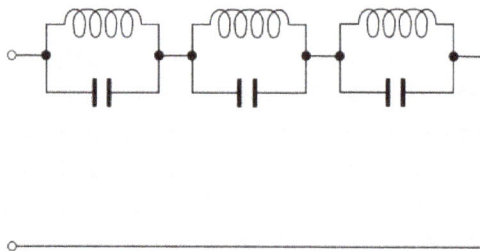

Foster's first form of canonical driving point impedance realisation. If the polynomial function has a pole at $\omega=0$ one of the LC sections will reduce to a single capacitor. If the polynomial function has a pole at $\omega=\infty$ one of the LC sections will reduce to a single inductor. If both poles are present then two sections reduce to a series LC circuit.

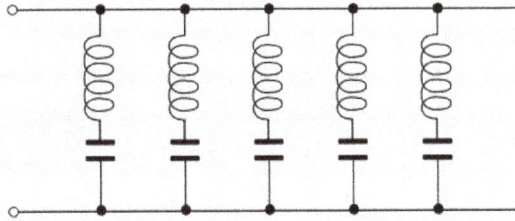

Foster's second form of canonical driving point impedance realisation. If the polynomial function has a zero at $\omega=0$ one of the LC sections will reduce to a single capacitor. If the polynomial function has a zero at $\omega=\infty$ one of the LC sections will reduce to a single inductor. If both zeroes are present then two sections reduce to a parallel LC circuit.

A one-port passive immittance consisting of discrete elements (that is, not a distributed element circuit) can be represented as a rational function of s,

$$Z(s) = \frac{P(s)}{Q(s)}$$

where,

$Z(s)$ is immittance

$P(s), Q(s)$ are polynomials with real, positive coefficiencts

s is the Laplace transform variable, which can be replaced with $i\omega$ when dealing with steady-state AC signals.

This follows from the fact the impedance of L and C elements are themselves simple rational functions and any algebraic combination of rational functions results in another rational function.

This is sometimes referred to as the driving point impedance because it is the impedance at the place in the network at which the external circuit is connected and "drives" it with a signal. In his paper, Foster describes how such a lossless rational function may be realised (if it can be realised) in two ways. Foster's first form consists of a number of series connected parallel LC circuits. Foster's second form of driving point impedance consists of a number of parallel connected series LC circuits. The realisation of the driving point impedance is by no means unique. Foster's realisation has the advantage that the poles and/or zeroes are directly associated with a particular resonant circuit, but there are many other realisations. Perhaps the most well known is Cauer's ladder realisation from filter design.

Non-Foster Networks

A Foster network must be passive, so an active network, containing a power source, may not obey Foster's theorem. These are called non-Foster networks. In particular, circuits containing an amplifier with positive feedback can have reactance which declines with frequency. For example, it is possible to create negative capacitance and inductance with negative impedance converter circuits. These circuits will have an immittance function with a phase of $\pm\pi/2$ like a positive reactance but a reactance amplitude with a negative slope against frequency.

These are of interest because they can accomplish tasks a Foster network cannot. For example, the usual passive Foster impedance matching networks can only match the impedance of an antenna with a transmission line at discrete frequencies, which limits the bandwidth of the antenna. A non-Foster network could match an antenna over a continuous band of frequencies. This would allow the creation of compact antennas that have wide bandwidth, violating the Chu-Harrington limit. Practical non-Foster networks are an active area of research.

History

The theorem was developed at American Telephone & Telegraph as part of ongoing investigations into improved filters for telephone multiplexing applications. This work was commercially important; large sums of money could be saved by increasing the number of telephone conversations that could be carried on one line. The theorem was first published by Campbell in 1922 but without a proof. Great use was immediately made of the theorem in filter design, it appears prominently, along with a proof, in Zobel's landmark paper of 1923 which summarised the state of the art of filter filter design at that time. Foster published his paper the following year which included his canonical realisation forms.

Cauer in Germany grasped the importance of Foster's work and used it as the foundation of network synthesis. Amongst Cauer's many innovations was the extension of Foster's work to all 2-element-kind networks after discovering an isomorphism between them. Cauer was interested in finding the necessary and sufficient condition for realisability of a rational one-port network from its polynomial function, a condition now known to be a positive-real function, and the reverse problem of which networks were equivalent, that is, had the same polynomial function. Both of these were important problems in network theory and filter design. Foster networks are only a subset of realisable networks,

Belevitch's Theorem

Belevitch's theorem is a theorem in electrical network analysis due to the Russo-Belgian mathematician Vitold Belevitch (1921–1999). The theorem provides a test for a given S-matrix to determine whether or not it can be constructed as a lossless rational two-port network.

Lossless implies that the network contains only inductances and capacitances - no resistances. Rational (meaning the driving point impedance $Z(p)$ is a rational function of p) implies that the network consists solely of discrete elements (inductors and capacitors only - no distributed elements).

The Theorem

For a given S-matrix $\mathbf{S}(p)$ of degree d;

$$\mathbf{S}(p) = \begin{bmatrix} S_{11} & S_{12} \\ S_{21} & S_{22} \end{bmatrix}$$

where,

p is the complex frequency variable and may be replaced by $i\omega$ in the case of steady state sine wave signals, that is, where only a Fourier analysis is required

d will equate to the number of elements (inductors and capacitors) in the network, if such network exists.

Belevitch's theorem states that, $\mathbf{S}(p)$ represents a lossless rational network if and only if,

$$\mathbf{S}(p) = \frac{1}{g(p)} \begin{bmatrix} h(p) & f(p) \\ \pm f(-p) & \mp h(-p) \end{bmatrix}$$

where,

$f(p)$, $g(p)$ and $h(p)$ are real polynomials

$g(p)$ is a strict Hurwitz polynomial of degree not exceeding

$g(p)g(-p) = f(p)f(-p) + h(p)h(-p)$ for all $p \in \mathbb{C}$.

Extra Element Theorem

The Extra Element Theorem (EET) is an analytic technique developed by R. D. Middlebrook for simplifying the process of deriving driving point and transfer functions for linear electronic circuits. Much like Thévenin's theorem, the extra element theorem breaks down one complicated problem into several simpler ones.

Driving point and transfer functions can generally be found using KVL and KCL methods, however several complicated equations may result that offer little insight into the circuit's behavior. Using the extra element theorem, a circuit element (such as a resistor) can be removed from a circuit and the desired driving point or transfer function found. By removing the element that most complicates the circuit (such as an element that creates feedback), the desired function can be easier to obtain. Next two correctional factors must be found and combined with the previously derived function to find the exact expression.

The general form of the extra element theorem is called the N-extra element theorem and allows multiple circuit elements to be removed at once.

General Formulation

The (single) extra element theorem expresses any transfer function as a product of the transfer function with that element removed and a correction factor. The correction factor term consists of the impedance of the extra element and two driving point impedances seen by the extra element: The double null injection driving point impedance and the single injection driving point impedance. Because an extra element can be removed in general by either short-circuiting or open-circuiting the element, there are two equivalent forms of the EET:

$$H(s) = H_\infty(s) \frac{1 + \dfrac{Z_n(s)}{Z(s)}}{1 + \dfrac{Z_d(s)}{Z(s)}}$$

or,

$$H(s) = H_0(s) \frac{1 + \dfrac{Z(s)}{Z_n(s)}}{1 + \dfrac{Z(s)}{Z_d(s)}}.$$

Where the Laplace-domain transfer functions and impedances in the above expressions are defined as follows: $H(s)$ is the transfer function with the extra element present. $H_\infty(s)$ is the transfer function with the extra element open-circuited. $H_o(s)$ is the transfer function with the extra element short-circuited. $Z(s)$ is the impedance of the extra element. $Z_d(s)$ is the single-injection driving point impedance "seen" by the extra element. $Z_n(s)$ is the double-null-injection driving point impedance "seen" by the extra element.

Driving Point Impedances

Single Injection Driving Point Impedance

$Z_d(s)$ is found by making the input to the system's transfer function zero (short circuit a voltage source or open circuit a current source) and determining the impedance across the terminals to which the extra element will be connected with the extra element absent.This impedance is same as the Thévenin's equivalent impedance.

Double Null Injection Driving Point Impedance

$Z_n(s)$ is found by replacing the extra element with a second test signal source (either current source or voltage source as appropriate). Then, $Z_n(s)$ is defined as the ratio of voltage across the terminals of this second test source to the current leaving its positive terminal when the output of the system's transfer function is nulled for any value of the primary input to the system's transfer function.

In practice, $Z_n(s)$ can be found from working backwards from the facts that the output of the transfer function is made zero and that the primary input to the transfer function is unknown. Then using conventional circuit analysis techniques to express both the voltage across the extra element test source's terminals, $v_n(s)$, and the current leaving the extra element test source's positive terminals, $i_n(s)$, and calculating $Z_n(s) = v_n(s)/i_n(s)$. Although computation of $Z_n(s)$ is an unfamiliar process for many engineers, its expressions are often much simpler than those for $Z_d(s)$ because the nulling of the transfer function's output often leads to other voltages/currents in the circuit being zero, which may allow exclusion of certain components from analysis.

Special Case With Transfer Function as a Self-Impedance

As a special case, the EET can be used to find the input impedance of a network with the addition of an element designated as "extra". In this case, Z_d is same as the impedance of the input test current source signal made zero or equivalently with the input open circuited. Likewise, since the transfer function output signal can be considered to be the voltage at the input terminals, Z_n is found when the input voltage is zero i.e. the input terminals are short-circuited. Thus, for this particular application the EET can be written as:

$$Z_{in} = Z_{in}^{\infty} \cdot \frac{1 + \dfrac{Z_e^0}{Z}}{1 + \dfrac{Z_e^{\infty}}{Z}}$$

where

Z is the impedance chosen as the extra element

Z_{in}^{∞} is the input impedance with Z removed (or made infinite)

Z_e^0 is the impedance seen by the extra element Z with the input shorted (or made zero)

Z_e^{∞} is the impedance seen by the extra element Z with the input open (or made infinite)

Computing these three terms may seem like extra effort, but they are often easier to compute than the overall input impedance.

Example

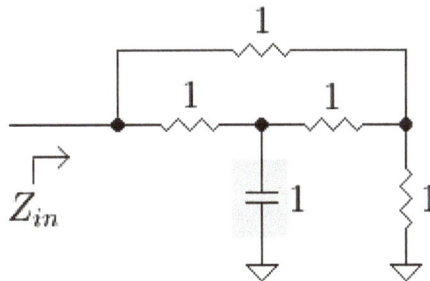

Figure 1: Simple RC circuit to demonstrate the EET. The capacitor (gray shading) is denoted the extra element

Consider the problem of finding Z_{in} for the circuit in Figure 1 using the EET (note all component values are unity for simplicity). If the capacitor (gray shading) is denoted the extra element then

$$Z = \frac{1}{s}$$

Removing this capacitor from the circuit we find

$$Z_{in}^{\infty} = 2\backslash 1 + 1 = \frac{5}{3}$$

Calculating the impedance seen by the capacitor with the input shorted we find

$$Z_e^0 = 1\backslash(1+1\backslash 1) = \frac{3}{5}$$

Calculating the impedance seen by the capacitor with the input open we find

$$Z_e^{\infty} = 2\backslash 1 + 1 = \frac{5}{3}$$

Therefore using the EET, we find

$$Z_{in} = \frac{5}{3} \cdot \frac{1+\dfrac{3}{5}s}{1+\dfrac{5}{3}s} = \frac{5+3s}{3+5s}$$

Note that this problem was solved by calculating three simple driving point impedances by inspection.

Feedback Amplifiers

The EET is also useful for analyzing single and multi-loop feedback amplifiers. In this case the EET can take the form of the Asymptotic gain model.

Commensurate Line circuit

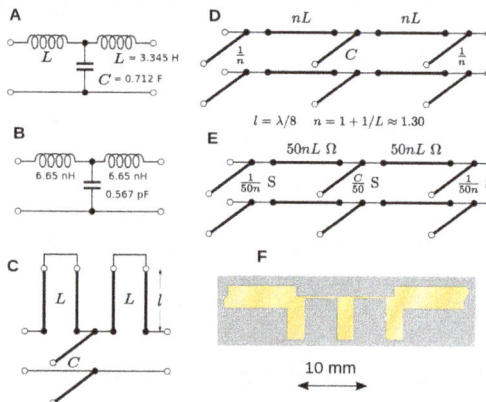

Example commensurate line design for a 4 GHz, 50 Ω, third order 3 dB Chebyshev low-pass filter. A. Prototype filter in lumped elements, $\omega=1$, $Z_0=1$. B. Filter frequency and impedance scaled to 4 GHz and 50 Ω; these component values are too small to easily implement as discrete components. C. The prototype circuit transformed to open-wire commensurate lines by Richards' transformation. D. Applying Kuroda's identities to prototype to eliminate the series inductors. E. Impedance scaling for 50 Ω working, frequency scaling is achieved by setting the line lengths to $\lambda/8$. F. Implementation in microstrip.

Commensurate line circuits are electrical circuits composed of transmission lines that are all the same length; commonly one-eighth of a wavelength. Lumped element circuits can be directly converted to distributed element circuits of this form by the use of Richards' transformation. This transformation has a particularly simple result; inductors are replaced with transmission lines terminated in short-circuits and capacitors are replaced with lines terminated in open-circuits. Commensurate line theory is particularly useful for designing distributed element filters for use at microwave frequencies.

It is usually necessary to carry out a further transformation of the circuit using Kuroda's identities. There are several reasons for applying one of the Kuroda transformations; the principal reason is usually to eliminate series connected components. In some technologies, including the widely used microstrip, series connections are difficult or impossible to implement.

The frequency response of commensurate line circuits, like all distributed element circuits, will periodically repeat, limiting the frequency range over which they are effective. Circuits designed by the methods of Richards and Kuroda are not the most compact. Refinements to the methods of coupling elements together can produce more compact designs. Nevertheless, the commensurate line theory remains the basis for many of these more advanced filter designs.

Commensurate Lines

Commensurate lines are transmission lines that are all the same electrical length, but not necessarily the same characteristic impedance (Z_0). A commensurate line circuit is an electrical circuit composed only of commensurate lines terminated with resistors or short- and open-circuits. In 1948, Paul I. Richards published a theory of commensurate line circuits by which a passive lumped element circuit could be transformed into a distributed element circuit with precisely the same characteristics over a certain frequency range.

Lengths of lines in distributed element circuits, for generality, are usually expressed in terms of the circuit's nominal operational wavelength, λ. Lines of the prescribed length in a commensurate line circuit are called *unit elements* (UEs). A particularly simple relationship pertains if the UEs are $\lambda/8$. Each element in the lumped circuit is transformed into a corresponding UE. However, Z_0 of the lines must be set according to the component value in the analogous lumped circuit and this may result in values of Z_0 that are not practical to implement. This is particularly a problem with printed technologies, such as microstrip, when implementing high characteristic impedances. High impedance requires narrow lines and there is a minimum size that can be printed. Very wide lines, on the other hand, allow the possibility of undesirable transverse resonant modes to form. A different length of UE, with a different Z_0, may be chosen to overcome these problems.

Electrical length can also be expressed as the phase change between the start and the end of the line. Phase is measured in angle units. θ, the mathematical symbol for an angle variable, is used as the symbol for electrical length when expressed as an angle. In this convention λ represents 360°, or 2π radians.

The advantage of using commensurate lines is that the commensurate line theory allows circuits to be synthesised from a prescribed frequency function. While any circuit using arbitrary transmission line lengths can be analysed to determine its frequency function, that circuit cannot necessarily be easily synthesised starting from the frequency function. The fundamental problem is that using more than one length generally requires more than one frequency variable. Using commensurate lines requires only one frequency variable. A well developed theory exists for synthesising lumped element circuits from a given frequency function. Any circuit so synthesised can be converted to a commensurate line circuit using Richards' transformation and a new frequency variable.

Richards' Transformation

Richards' transformation transforms the angular frequency variable, ω, according to,

$$\omega \rightarrow \tan(k\omega)$$

or, more usefully for further analysis, in terms of the complex frequency variable, s,

$$s \rightarrow \tanh(ks)$$

where k is an arbitrary constant related to the UE length, θ, and some designer chosen reference frequency, ω_c, by

$$k\omega_c = \theta.$$

k has units of time and is, in fact, the phase delay inserted by a UE.

Comparing this transform with expressions for the driving point impedance of stubs terminated, respectively, with a short circuit and an open circuit,

$$Z_{SC} = jZ_0 \tan(k\omega)$$
$$Z_{OC} = -jZ_0 \cot(k\omega)$$

it can be seen that (for $\theta < \pi/2$) a short circuit stub has the impedance of a lumped inductance and an open circuit stub has the impedance of a lumped capacitance. Richards' transformation substitutes inductors with short circuited UEs and capacitors with open circuited UEs.

When the length is $\lambda/8$ (or $\theta=\pi/4$), this simplifies to,

$$Z_{SC} = jZ_0$$
$$Z_{OC} = -jZ_0$$

This is frequently written as,

$$Z_{SC} = jL$$
$$Z_{OC} = \frac{1}{jC}$$

L and C are conventionally the symbols for inductance and capacitance, but here they represent respectively the characteristic impedance of an inductive stub and the characteristic admittance of a capacitive stub. This convention is used by numerous authors, and later in this article.

Omega-domain

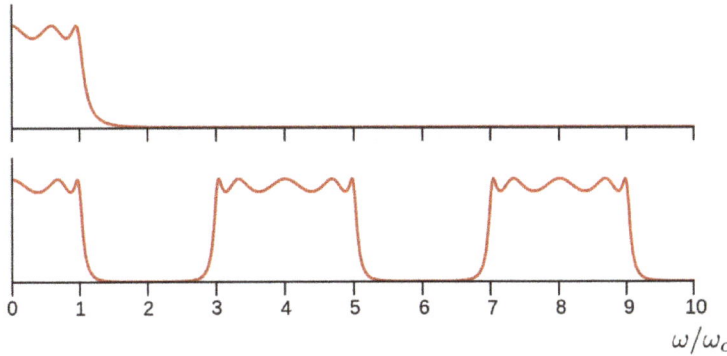

Frequency response of a fifth order Chebyshev filter (top), and the same filter after applying Richards' transformation

Richards' transformation can be viewed as transforming from a s-domain representation to a new domain called the Ω-domain where,

$$\Omega = \tan(k\omega)$$

If Ω is normalised so that $\Omega=1$ when $\omega=\omega_c$, then it is required that,

$$k\omega_c = \theta = \frac{\pi}{4}$$

and the length in distance units becomes,

$$\ell = \frac{\lambda}{8}$$

Any circuit composed of discrete, linear, lumped components will have a transfer function $H(s)$ that is a rational function in s. A circuit composed of transmission line UEs derived from the lumped circuit by Richards' transformation will have a transfer function $H(j\Omega)$ that is a rational function of precisely the same form as $H(s)$. That is, the shape of the frequency response of the lumped circuit against the s frequency variable will be precisely the same as the shape of the frequency response of the transmission line circuit against the $j\Omega$ frequency variable and the circuit will be functionally the same.

However, infinity in the Ω domain is transformed to $\omega=\pi/4k$ in the s domain. The entire frequency response is squeezed down to this finite interval. Above this frequency, the same response is repeated in the same intervals, alternately in reverse. This is a consequence of the periodic nature of the tangent function. This multiple passband result is a general feature of all distributed element circuits, not just those arrived at through Richards' transformation.

Cascade Element

A UE connected in cascade is a two-port network that has no exactly corresponding circuit in lumped elements. It is functionally a fixed delay. There are lumped-element circuits that can approximate a fixed delay such as the Bessel filter, but they only work within a prescribed passband, even with ideal components. Alternatively, lumped-element all-pass filters can be constructed that pass all frequencies (with ideal components), but they have constant delay only within a narrow band of frequencies. Examples are the lattice phase equaliser and bridged T delay equaliser.

There is consequently no lumped circuit that Richard's transformation can transform into a cascade-connected line, and there is no reverse transformation for this element. Commensurate line theory thus introduces a new element of *delay*, or *length*. Two or more UEs connected in cascade with the same Z_0 are equivalent to a single, longer, transmission line. Thus, lines of length $n\theta$ for integer n are allowable in commensurate circuits. Some circuits can be implemented *entirely* as a cascade of UEs: impedance matching networks, for instance, can be done this way, as can most filters.

Kuroda's Identities

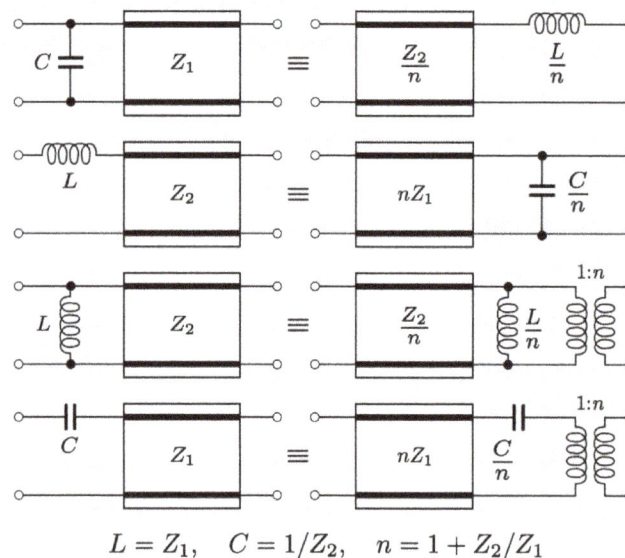

$$L = Z_1, \quad C = 1/Z_2, \quad n = 1 + Z_2/Z_1$$

Kuroda's identities
Kuroda's identities are a set of four equivalent circuits that overcome certain difficulties with applying Richards' transformations directly. The four basic transformations are shown in the figure. Here the symbols for capacitors and inductors are used to represent open-circuit and short-circuit stubs. Likewise, the symbols C and L here represent respectively the susceptance of an open circuit stub and the reactance of a short circuit stub, which, for $\theta=\lambda/8$, are respectively equal to the characteristic admittance and characteristic impedance of the stub line. The boxes with thick lines represent cascade connected commensurate lengths of line with the marked characteristic impedance.

The first difficulty solved is that all the UEs are required to be connected together at the same point. This arises because the lumped element model assumes that all the elements take up zero space (or no significant space) and that there is no delay in signals between the elements. Applying Richards' transformation to convert the lumped circuit into a distributed circuit allows the element to now occupy a finite space (its length) but does not remove the requirement for zero distance between the interconnections. By repeatedly applying the first two Kuroda identities, UE lengths of the lines feeding into the ports of the circuit can be moved between the circuit components to physically separate them.

A second difficulty that Kuroda's identities can overcome is that series connected lines are not always practical. While series connection of lines can easily be done in, for instance, coaxial technology, it is not possible in the widely used microstrip technology and other planar technologies. Filter circuits frequently use a ladder topology with alternating series and shunt elements. Such circuits can be converted to all shunt components in the same step used to space the components with the first two identities.

The third and fourth identities allow characteristic impedances to be scaled down or up respectively. These can be useful for transforming impedances that are impractical to implement. However, they have the disadvantage of requiring the addition of an ideal transformer with a turns ratio equal to the scaling factor.

History

In the decade after Richards' publication, advances in the theory of distributed circuits took place mostly in Japan. K. Kuroda published these identities in 1955 in his Ph.D thesis. However, they did not appear in English until 1958 in a paper by Ozaki and Ishii on stripline filters.

Further Refinements

One of the major applications of commensurate line theory is to design distributed element filters. Such filters constructed directly by Richards' and Kuroda's method are not very compact. This can be an important design consideration, especially in mobile devices. The stubs stick out to the side of the main line and the space between them is not doing anything useful. Ideally, the stubs should project on alternate sides to prevent them coupling with each other, taking up further space, although this is not always done for space considerations. More than that, the cascade connected elements that couple together the stubs contribute nothing to the frequency function, they are only there to transform the stubs into the required impedance. Putting it another way, the order of the frequency function is determined solely by the number of stubs, not by the total number of UEs (generally speaking, the higher the order, the better the filter). More complex synthesis techniques can produce filters in which all elements are contributing.

The cascade connected $\lambda/8$ sections of the Kuroda circuits are an example of impedance transformers, the archetypical example of such circuits is the $\lambda/4$ impedance transformer. Although this is double the length of the $\lambda/8$ line it has the useful property that it can be transformed from a low-pass filter to a high-pass filter by replacing the open circuit stubs with short circuit stubs. The two filters are exactly matched with the same cut-off frequency and mirror-symmetrical responses. It is therefore ideal for use in diplexers. The $\lambda/4$ transformer has this property of being invariant under a low-pass to high-pass transformation because it is not just an impedance transformer, but a special case of transformer, an impedance inverter. That is, it transforms any impedance network at one port, to the inverse impedance, or dual impedance, at the other port. However, a single length of transmission line can only be precisely $\lambda/4$ long at its resonant frequency and there is consequently a limit to the bandwidth over which it will work. There are more complex kinds of inverter circuit that more accurately invert impedances. There are two classes of inverter, the J-inverter, which transforms a shunt admittance into a series impedance, and the K-inverter which does the reverse transformation. The coefficients J and K are respectively the scaling admittance and impedance of the converter.

Stubs may be lengthened in order to change from an open circuit to a short circuit stub and vice versa. Low-pass filters usually consist of series inductors and shunt capacitors. Applying Kuroda's identities will convert these to all shunt capacitors, which are open circuit stubs. Open circuit stubs are preferred in printed technologies because they are easier to implement, and this is the technology likely to be found in consumer products. However, this is not the case in other technologies such as coaxial line, or twin-lead where the short circuit may actually be helpful for mechanical support of the structure. Short circuits also have a small advantage in that they are generally have a more precise position than open circuits. If the circuit is to be further transformed into the waveguide medium then open circuits are out of the question because there would be radiation out of the aperture so formed. For a high-pass filter the inverse applies, applying Kuroda will naturally result in short circuit stubs and it may be desirable for a printed design to convert to open circuits. As an example, a $\lambda/8$ open circuit stub can be replaced with a $3\lambda/8$ short circuit stub of the same characteristic impedance without changing the circuit functionally.

Coupling elements together with impedance transformer lines is not the most compact design. Other methods of coupling have been developed, especially for band-pass filters that are far more compact. These include parallel lines filters, interdigital filters, hairpin filters, and the semi-lumped design combline filters.

Equivalent Impedance Transforms

An equivalent impedance is an equivalent circuit of an electrical network of impedance elements which presents the same impedance between all pairs of terminals as did the given network. This article describes mathematical transformations between some passive, linear impedance networks commonly found in electronic circuits.

There are a number of very well known and often used equivalent circuits in linear network analysis. These include resistors in series, resistors in parallel and the extension to series and parallel circuits for capacitors, inductors and general impedances. Also well known are the Norton and Thévenin equivalent current generator and voltage generator circuits respectively, as is the Y-Δ transform. None of these are discussed in detail here; the individual linked articles should be consulted.

The number of equivalent circuits that a linear network can be transformed into is unbounded. Even in the most trivial cases this can be seen to be true, for instance, by asking how many different combinations of resistors in parallel are equivalent to a given combined resistor. The number of series and parallel combinations that can be formed grows exponentially with the number of resistors, n. For large n the size of the set has been found by numerical techniques to be approximately 2.53^n and analytically strict bounds are given by a Farey sequence of Fibonacci numbers. This article could never hope to be comprehensive, but there are some generalisations possible. Wilhelm Cauer found a transformation that could generate all possible equivalents of a given rational, passive, linear one-port, or in other words, any given two-terminal impedance. Transformations of 4-terminal, especially 2-port, networks are also commonly found and transformations of yet more complex networks are possible.

The vast scale of the topic of equivalent circuits is underscored in a story told by Sidney Darlington. According to Darlington, a large number of equivalent circuits were found by Ronald Foster, following his and George Campbell's 1920 paper on non-dissipative four-ports. In the course of this work they looked at the ways four ports could be interconnected with ideal transformers[note 5] and maximum power transfer. They found a number of combinations which might have practical applications and asked the AT&T patent department to have them patented. The patent department replied that it was pointless just patenting some of the circuits if a competitor could use an equivalent circuit to get around the patent; they should patent all of them or not bother. Foster therefore set to work calculating every last one of them. He arrived at an enormous total of 83,539 equivalents (577,722 if different output ratios are included). This was too many to patent, so instead the information was released into the public domain in order to prevent any of AT&T's competitors from patenting them in the future.

2-terminal, 2-element-kind Networks

A single impedance has two terminals to connect to the outside world, hence can be described as a 2-terminal, or a one-port, network. Despite the simple description, there is no limit to the number of meshes, and hence complexity and number of elements, that the impedance network may have. 2-element-kind networks are common in circuit design; filters, for instance, are often LC-kind networks and printed circuit designers favour RC-kind networks because inductors are less easy to manufacture. Transformations are simpler and easier to find than for 3-element-kind networks. One-element-kind networks can be thought of as a special case of two-element-kind. It is possible to use the transformations in this section on a certain few 3-element-kind networks by substituting a network of elements for element Z_n. However, this is limited to a maximum of two impedances being substituted; the remainder will not be a free choice. All the transformation equations given in this section are due to Otto Zobel.

3-element Networks

One-element networks are trivial and two-element, two-terminal networks are either two elements in series or two elements in parallel, also trivial. The smallest number of elements that is non-trivial is three, and there are two 2-element-kind non-trivial transformations possible, one being both the reverse transformation and the topological dual, of the other.

Description	Network	Transform equations	Transformed network
Transform 1.1 Transform 1.2 is the reverse of this transform.	Z_1, $m_1 Z_1$, Z_2	$p_1 = 1 + m_1$, $p_2 = m_1(1 + m_1)$, $p_3 = (1 + m_1)^2$.	$p_1 Z_1$, $p_2 Z_1$, $p_3 Z_2$
Transform 1.2 The reverse transform, and topological dual, of Transform 1.1.	$m_1 Z_1$, Z_1, Z_2	$p_1 = \dfrac{m_1^2}{1 + m_1}$, $p_2 = \dfrac{m_1}{1 + m_1}$, $p_3 = \left(\dfrac{m_1}{1 + m_1}\right)^2$.	$p_1 Z_1$, $p_2 Z_1$, $p_3 Z_2$

Example 1. An example of Transform 1.2. The reduced size of the inductor has practical advantages.		$m_1 = 0.5$, $p_1 = \dfrac{1}{6}$, $p_2 = \dfrac{1}{3}$, $p_3 = \dfrac{1}{9}$.	

4-element Networks

There are four non-trivial 4-element transformations for 2-element-kind networks. Two of these are the reverse transformations of the other two and two are the dual of a different two. Further transformations are possible in the special case of Z_2 being made the same element kind as Z_1, that is, when the network is reduced to one-element-kind. The number of possible networks continues to grow as the number of elements is increased. For all entries in the following table it is defined:

$$q_1 := 1 + m_1 + m_2,$$
$$q_2 := \sqrt{q_1^2 - 4m_1m_2},$$
$$q_3 := \frac{(1+m_1)(1+m_2)}{(m_1 - m_2)^2},$$

$$q_4 := \frac{q_2 - q_1 + 2m_2}{2q_2},$$
$$q_5 := \frac{q_2 + q_1 - 2m_2}{2q_2}.$$

Description	Network	Transform equations	Transformed network
Transform 2.1 Transform 2.2 is the reverse of this transform. Transform 2.3 is the topological dual of this transform.		$p_1 = \dfrac{q_1 + q_2}{2q_5}$, $p_2 = \dfrac{q_1 - q_2}{2q_4}$, $p_3 = \dfrac{m_2}{q_5}$, $p_4 = \dfrac{m_2}{q_4}.$	
Transform 2.2 Transform 2.1 is the reverse of this transform. Transform 2.4 is the topological dual of this transform.		$p_1 = \dfrac{1}{q_3(1+m_2)}$, $p_2 = \dfrac{m_1}{1+m_1}$, $p_3 = \dfrac{1}{q_3(1+m_1)}$, $p_4 = \dfrac{m_2}{1+m_2}.$	

Transform 2.3 Transform 2.4 is the reverse of this transform. Transform 2.1 is the topological dual of this transform.	Z_1 Z_2 $m_1 Z_1$ $m_2 Z_2$	$p_1 = \dfrac{q_4(q_1 + q_2)}{2m_2}$, $p_2 = \dfrac{q_5(q_1 - q_2)}{2m_2}$, $p_3 = q_4$, $p_4 = q_5$.	$p_1 Z_1$ $p_2 Z_1$ $p_3 Z_2$ $p_4 Z_2$
Transform 2.4 Transform 2.3 is the reverse of this transform. Transform 2.2 is the topological dual of this transform.	Z_1 $m_1 Z_1$ Z_2 $m_2 Z_2$	$p_1 = 1 + m_1$, $p_2 = m_1 q_3 (1 + m_1)$, $p_3 = 1 + m_2$, $p_4 = m_1 q_3 (1 + m_2)$.	$p_1 Z_1$ $p_3 Z_2$ $p_2 Z_1$ $p_4 Z_2$
Example 2. An example of Transform 2.2.	$3R_1$ C_2 R_1 C_2	$m_1 = 3$, $m_2 = 1$, $q_3 = 2$, $p_1 = \dfrac{1}{4}$, $p_2 = \dfrac{3}{4}$, $p_3 = \dfrac{1}{8}$, $p_4 = \dfrac{1}{2}$.	$\frac{1}{4}R_1$ $\frac{3}{4}R_1$ $2C_2$ $8C_2$

2-terminal, *n*-element, 3-element-kind Networks

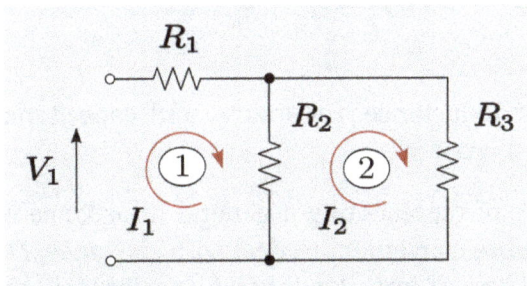

Fig. 1. Simple example of a network of impedances using resistors only for clarity. However, analysis of networks with other impedance elements proceed by the same principles. Two meshes are shown, with numbers in circles. The sum of impedances around each mesh, p, will form the diagonal of the entries of the matrix, Z_{pp}. The impedance of branches shared by two meshes, p and q, will form the entries $-Z_{pq}$. Z_{pq}, p≠q, will always have a minus sign provided that the convention of loop currents are defined in the same (conventionally counter-clockwise) direction and the mesh contains no ideal transformers or mutual inductors.

Simple networks with just a few elements can be dealt with by formulating the network equations "by hand" with the application of simple network theorems such as Kirchhoff's laws. Equivalence is proved between two networks by directly comparing the two sets of equations and equating coefficients. For large networks more powerful techniques are required. A common approach is to start by expressing the network of impedances as a matrix. This approach is only good for rational[note 9] networks. Any network that includes distributed elements, such as a transmission line, cannot be represented by a finite matrix. Generally, an n-mesh[note 6] network requires an nxn matrix to represent it. For instance the matrix for a 3-mesh network might look like

$$[\mathbf{Z}] = \begin{bmatrix} Z_{11} & Z_{12} & Z_{13} \\ Z_{21} & Z_{22} & Z_{23} \\ Z_{31} & Z_{32} & Z_{33} \end{bmatrix}$$

The entries of the matrix are chosen so that the matrix forms a system of linear equations in the mesh voltages and currents (as defined for mesh analysis):

$$[\mathbf{V}] = [\mathbf{Z}][\mathbf{I}]$$

The example diagram in Figure 1, for instance, can be represented as an impedance matrix by

$$[\mathbf{Z}] = \begin{bmatrix} R_1 + R_2 & -R_2 \\ -R_2 & R_2 + R_3 \end{bmatrix}$$

and the associated system of linear equations is

$$\begin{bmatrix} V_1 \\ 0 \end{bmatrix} = \begin{bmatrix} R_1 + R_2 & -R_2 \\ -R_2 & R_2 + R_3 \end{bmatrix} \begin{bmatrix} I_1 \\ I_2 \end{bmatrix}$$

In the most general case, each branch[note 1] Z_p of the network may be made up of three elements so that

$$Z_p = sL_p + R_p + \frac{1}{sC_p}$$

where L, R and C represent inductance, resistance, and capacitance respectively and s is the complex frequency operator $s = \sigma + i\omega$.

This is the conventional way of representing a general impedance but for the purposes of this article it is mathematically more convenient to deal with elastance, D, the inverse of capacitance, C. In those terms the general branch impedance can be represented by

$$sZ_p = s^2 L_p + sR_p + D_p$$

Likewise, each entry of the impedance matrix can consist of the sum of three elements. Consequently, the matrix can be decomposed into three nxn matrices, one for each of the three element kinds:

$$s[\mathbf{Z}] = s^2[\mathbf{L}] + s[\mathbf{R}] + [\mathbf{D}]$$

It is desired that the matrix [Z] represent an impedance, $Z(s)$. For this purpose, the loop of one of the meshes is cut and $Z(s)$ is the impedance measured between the points so cut. It is conventional to assume the external connection port is in mesh 1, and is therefore connected across matrix entry Z_{11}, although it would be perfectly possible to formulate this with connections to any desired nodes. In the following discussion $Z(s)$ taken across Z_{11} is assumed. $Z(s)$ may be calculated from [Z] by

$$Z(s) = \frac{|\mathbf{Z}|}{z_{11}}$$

where z_{11} is the complement of Z_{11} and $|\mathbf{Z}|$ is the determinant of [Z].

For the example network above,

$$|\mathbf{Z}| = (R_1 + R_2)(R_2 + R_3) - R_2^2 = R_1 R_2 + R_1 R_3 + R_2 R_3 ,$$

$$z_{11} = Z_{22} = R_2 + R_3 , \text{ and,}$$

$$Z(s) = R_1 + \frac{R_2 R_3}{R_2 + R_3} .$$

This result is easily verified to be correct by the more direct method of resistors in series and parallel. However, such methods rapidly become tedious and cumbersome with the growth of the size and complexity of the network under analysis.

The entries of [R], [L] and [D] cannot be set arbitrarily. For [Z] to be able to realise the impedance $Z(s)$ then [R],[L] and [D] must all be positive-definite matrices. Even then, the realisation of $Z(s)$ will, in general, contain ideal transformers within the network. Finding only those transforms that do not require mutual inductances or ideal transformers is a more difficult task. Similarly, if starting from the "other end" and specifying an expression for $Z(s)$, this again cannot be done arbitrarily. To be realisable as a rational impedance, $Z(s)$ must be positive-real. The positive-real (PR) condition is both necessary and sufficient but there may be practical reasons for rejecting some topologies.

A general impedance transform for finding equivalent rational one-ports from a given instance of [Z] is due to Wilhelm Cauer. The group of real affine transformations

$$[\mathbf{Z}'] = [\mathbf{T}]^T [\mathbf{Z}][\mathbf{T}]$$

where

$$[\mathbf{T}] = \begin{bmatrix} 1 & 0 \cdots 0 \\ T_{21} & T_{22} \cdots T_{2n} \\ . & \cdots \\ T_{n1} & T_{n2} \cdots T_{nn} \end{bmatrix}$$

is invariant in $Z(s)$. That is, all the transformed networks are equivalents according to the definition given here. If the $Z(s)$ for the initial given matrix is realisable, that is, it meets the PR condition, then all the transformed networks produced by this transformation will also meet the PR condition.

3 and 4-terminal Networks

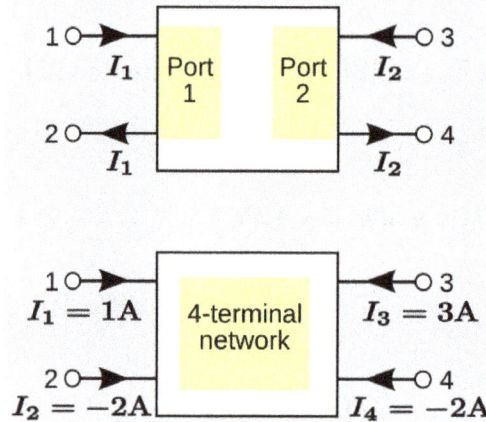

Fig. 2. A 4-terminal network connected by ports (top) has equal and opposite currents in each pair of terminals. The bottom network does not meet the port condition and cannot be treated as a 2-port. It could, however, be treated as an unbalanced 3-port by splitting one of the terminals into three common terminals shared between the ports.

When discussing 4-terminal networks, network analysis often proceeds in terms of 2-port networks, which covers a vast array of practically useful circuits. "2-port", in essence, refers to the way the network has been connected to the outside world: that the terminals have been connected in pairs to a source or load. It is possible to take exactly the same network and connect it to external circuitry in such a way that it is no longer behaving as a 2-port. This idea is demonstrated in Figure 2.

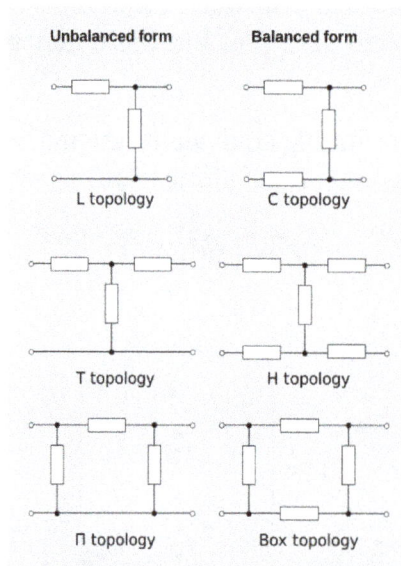

Equivalent unbalanced and balanced networks. The impedance of the series elements in the balanced version is half the corresponding impedance of the unbalanced version.

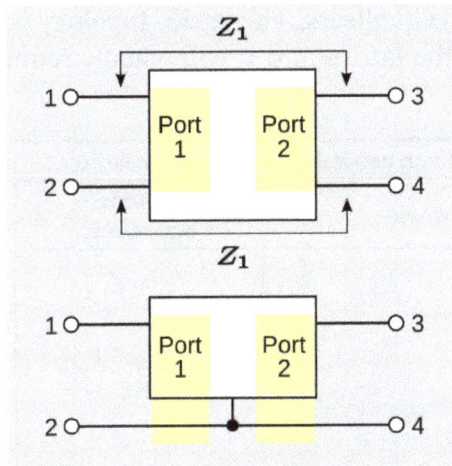

Fig. 3. To be balanced, a network must have the same impedance in each "leg" of the circuit.

A 3-terminal network can also be used as a 2-port. To achieve this, one of the terminals is connected in common to one terminal of both ports. In other words, one terminal has been split into two terminals and the network has effectively been converted to a 4-terminal network. This topology is known as unbalanced topology and is opposed to balanced topology. Balanced topology requires, referring to Figure 3, that the impedance measured between terminals 1 and 3 is equal to the impedance measured between 2 and 4. This is the pairs of terminals *not* forming ports: the case where the pairs of terminals forming ports have equal impedance is referred to as symmetrical. Strictly speaking, any network that does not meet the balance condition is unbalanced, but the term is most often referring to the 3-terminal topology described above and in Figure 3. Transforming an unbalanced 2-port network into a balanced network is usually quite straightforward: all series connected elements are divided in half with one half being relocated in what was the common branch. Transforming from balanced to unbalanced topology will often be possible with the reverse transformation but there are certain cases of certain topologies which cannot be transformed in this way. For example, see the discussion of lattice transforms below.

An example of a 3-terminal network transform that is not restricted to 2-ports is the Y-Δ transform. This is a particularly important transform for finding equivalent impedances. Its importance arises from the fact that the total impedance between two terminals cannot be determined solely by calculating series and parallel combinations except for a certain restricted class of network. In the general case additional transformations are required. The Y-Δ transform, its inverse the Δ-Y transform, and the *n*-terminal analogues of these two transforms (star-polygon transforms) represent the minimal additional transforms required to solve the general case. Series and parallel are, in fact, the 2-terminal versions of star and polygon topology. A common simple topology that cannot be solved by series and parallel combinations is the input impedance to a bridge network (except in the special case when the bridge is in balance). The rest of the transforms in this section are all restricted to use with 2-ports only.

Lattice Transforms

Symmetric 2-port networks can be transformed into lattice networks using Bartlett's bisection theorem. The method is limited to symmetric networks but this includes many topologies commonly found in filters, attenuators and equalisers. The lattice topology is intrinsically balanced, there is no unbalanced counterpart to the lattice and it will usually require more components than the transformed network.

Some common networks transformed to lattices (X-networks)			
Description	**Network**	**Transform equations**	**Transformed network**
Transform 3.1 Transform of T network to lattice network.		$Z_A = Z_1$, $Z_B = Z_1 + 2Z_2$.	
Transform 3.2 Transform of Π network to lattice network.		$Z_A = \dfrac{Z_1 Z_2}{Z_1 + 2Z_2}$, $Z_B = Z_2$.	
Transform 3.3 Transform of Bridged-T network to lattice network.		$Z_A = \dfrac{Z_1 Z_0}{Z_1 + 2Z_0}$, $Z_B = Z_0 + 2Z_2$.	

Reverse transformations from a lattice to an unbalanced topology are not always possible in terms of passive components. For instance, this transform,

Description	**Network**	**Transformed network**
Transform 3.4 Transform of a lattice phase equaliser to a T network.		

cannot be realised with passive components because of the negative values arising in the transformed circuit. It can however be realised if mutual inductances and ideal transformers are permitted, for instance, in this circuit. Another possibility is to permit the use of active components which would enable negative impedances to be directly realised as circuit components.

It can sometimes be useful to make such a transformation, not for the purposes of actually building the transformed circuit, but rather, for the purposes of aiding understanding of how the original circuit is working. The following circuit in bridged-T topology is a modification of a mid-series m-derived filter T-section. The circuit is due to Hendrik Bode who claims that the addition of the bridging resistor of a suitable value will cancel the parasitic resistance of the shunt inductor. The action of this circuit is clear if it is transformed into T topology - in this form there is a negative resistance in the shunt branch which can be made to be exactly equal to the positive parasitic resistance of the inductor.

Description	Network	Transformed network
Transform 3.5 Transform of a bridged-T low-pass filter section to a T-section.		

Any symmetrical network can be transformed into any other symmetrical network by the same method, that is, by first transforming into the intermediate lattice form (omitted for clarity from the above example transform) and from the lattice form into the required target form. As with the example, this will generally result in negative elements except in special cases.

Eliminating Resistors

A theorem due to Sidney Darlington states that any PR function $Z(s)$ can be realised as a lossless two-port terminated in a positive resistor R. That is, regardless of how many resistors feature in the matrix [Z] representing the impedance network, a transform can be found that will realise the network entirely as an LC-kind network with just one resistor across the output port (which would normally represent the load). No resistors within the network are necessary in order to realise the specified response. Consequently, it is always possible to reduce 3-element-kind 2-port networks to 2-element-kind (LC) 2-port networks provided the output port is terminated in a resistance of the required value.

Eliminating Ideal Transformers

An elementary transformation that can be done with ideal transformers and some other impedance element is to shift the impedance to the other side of the transformer. In all the following transforms, r is the turns ratio of the transformer.

Description	Network	Transformed network
Transform 4.1 Series impedance through a step-down transformer.		
Transform 4.2 Shunt impedance through a step-down transformer.		

| Transform 4.3 Shunt and series impedance network through a step-up transformer. | | |

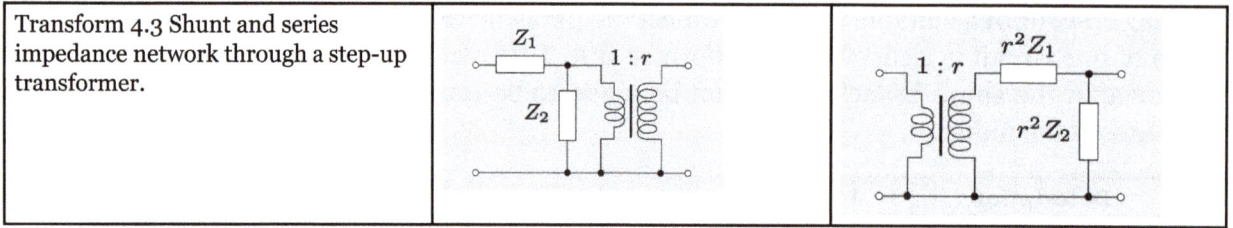

These transforms do not just apply to single elements; entire networks can be passed through the transformer. In this manner, the transformer can be shifted around the network to a more convenient location.

Darlington gives an equivalent transform that can eliminate an ideal transformer altogether. This technique requires that the transformer is next to (or capable of being moved next to) an "L" network of same-kind impedances. The transform in all variants results in the "L" network facing the opposite way, that is, topologically mirrored.

Description	Network	Transformed network
Transform 5.1 Elimination of a step-down transformer.		
Transform 5.2 Elimination of a step-up transformer.		
Example 3. Example of transform 5.1.		

Example 3 shows the result is a Π-network rather than an L-network. The reason for this is that the shunt element has more capacitance than is required by the transform so some is still left over after applying the transform. If the excess were instead, in the element nearest the transformer, this could be dealt with by first shifting the excess to the other side of the transformer before carrying out the transform.

Terminology

1. Branch. A network branch is a group of elements connected in series between two nodes. An essential feature of a branch is that all elements in the branch have the same current flowing through them.

2. Element. A component in a network, an individual resistor (R), inductor (L) or capacitor (C).

3. *n*-element. A network that contains a total of *n* elements of all kinds.

4. *n*-element-kind. A network that contains *n* different kinds of elements. For instance, a network consisting solely of LC elements is a 2-element-kind network.

5. Ideal transformer. These frequently appear in network analysis. They are a purely theoretical construct which perfectly transform voltages and currents by the given ratio without loss. Real transformers are highly efficient and can often be used in place of an ideal transformer. One essential difference is that ideal transformers continue to work when energised with DC, something no real transformer could ever do.

6. *n*-mesh. A mesh is a loop of a network where connections exist to allow current to pass from element to element, and form an unbroken path returning eventually to the starting point. An essential mesh is such a loop that does not contain any other loop. An *n*-mesh network is one that contains *n* essential meshes.

7. Node. A network node is point in a circuit where one terminal of three or more elements are joined.

8. Port. A pair of terminals of a network into which flows equal and opposite currents.

9. Rational in this context means a network composed of a finite number of elements. Distributed elements, such as in a transmission line, are therefore excluded because the infinitesimal nature of the elements will cause their number to go to infinity.

10. Terminal. A point in a network to which voltages external to the network can be connected and into which external currents may flow. A 2-terminal network is also a one-port network. 3-terminal and 4-terminal networks are often, but not always, also connected as 2-port networks.

Equivalent Circuit

In electrical engineering and science, an equivalent circuit refers to a theoretical circuit that retains all of the electrical characteristics of a given circuit. Often, an equivalent circuit is sought that simplifies calculation, and more broadly, that is a simplest form of a more complex circuit in order to aid analysis. In its most common form, an equivalent circuit is made up of linear, passive elements. However, more complex equivalent circuits are used that approximate the nonlinear behavior of the original circuit as well. These more complex circuits often are called macromodels of the original circuit. An example of a macromodel is the Boyle circuit for the 741 operational amplifier.

Equivalent circuits can also be used to electrically describe and model either a) continuous materials or biological systems in which current does not actually flow in defined circuits, or, b) distributed reactances, such as found in electrical lines or windings, that do not represent actual discrete components. For example, a cell membrane can be modelled as a capacitance (i.e. the lipid bilayer) in parallel with resistance-DC voltage source combinations (i.e. ion channels powered by an ion gradient across the membrane).

Examples

Thévenin and Norton Equivalents

One of linear circuit theory's most surprising properties relates to the ability to treat any two-terminal circuit no matter how complex as behaving as only a source and an impedance, which have either of two simple equivalent circuit forms:

- Thévenin equivalent - Any linear two-terminal circuit can be replaced by a single voltage source and a series impedance.

- Norton equivalent - Any linear two-terminal circuit can be replaced by a current source and a parallel impedance.

However, the single impedance can be of arbitrary complexity (as a function of frequency) and may be irreducible to a simpler form.

DC and AC Equivalent Circuits

In linear circuits, due to the superposition principle, the output of a circuit is equal to the sum of the output due to its DC sources alone, and the output from its AC sources alone. Therefore, the DC and AC response of a circuit is often analyzed independently, using separate DC and AC equivalent circuits which have the same response as the original circuit to DC and AC currents respectively. The composite response is calculated by adding the DC and AC responses:

- A DC equivalent of a circuit can be constructed by replacing all capacitances with open circuits, inductances with short circuits, and reducing AC sources to zero (replacing AC voltage sources by short circuits and AC current sources by open circuits.)

- An AC equivalent circuit can be constructed by reducing all DC sources to zero (replacing DC voltage sources with short circuits and DC current sources with open circuits)

This technique is often extended to small-signal nonlinear circuits like tube and transistor circuits, by linearizing the circuit about the DC bias point Q-point, using an AC equivalent circuit made by calculating the equivalent *small signal* AC resistance of the nonlinear components at the bias point.

Two-port Networks

Linear four-terminal circuits in which a signal is applied to one pair of terminals and an output is taken from another, are often modeled as two-port networks. These can be represented by simple equivalent circuits of impedances and dependent sources. To be analyzed as a two port network the currents applied to the circuit must satisfy the *port condition*: the current entering one terminal of a port must be equal to the current leaving the other terminal of the port. By linearizing a nonlinear circuit about its operating point, such a two-port representation can be made for transistors.

Delta and Wye Circuits

In three phase power circuits, three phase sources and loads can be connected in two different ways, called a "delta" connection and a "wye" connection. In analyzing circuits, sometimes it simplifies the analysis to convert between equivalent wye and delta circuits. This can be done with the wye-delta transform.

References

- Vorpérian, Vatché (2002). Fast analytical techniques for electrical and electronic circuits. Cambridge UK/NY: Cambridge University Press. pp. 61–106. ISBN 0-521-62442-8.

- Richard C. Dorf (1997). The Electrical Engineering Handbook. New York: CRC Press. Fig. 27.4, p. 711. ISBN 0-8493-8574-1.

- P.R. Gray; P.J. Hurst; S.H. Lewis; R.G. Meyer (2001). Analysis and Design of Analog Integrated Circuits (Fourth ed.). New York: Wiley. pp. §3.2, p. 172. ISBN 0-471-32168-0.

5

Understanding Electronics

Electronics is a field of science that has electricity as its most fundamental part. Some of the features of electronics elucidated in this text are digital electronics, analogue electronics, electronic circuit design and electronic components. The aspects elucidated in this section are of vital importance, and provides a better understanding of electronics.

Electronics

Electronics is the science of controlling electrical energy electrically, in which the electrons have a fundamental role. Electronics deals with electrical circuits that involve active electrical components such as vacuum tubes, transistors, diodes, integrated circuits, associated passive electrical components, and interconnection technologies. Commonly, electronic devices contain circuitry consisting primarily or exclusively of active semiconductors supplemented with passive elements; such a circuit is described as an electronic circuit.

Surface-mount electronic components

The science of Electronics is also considered to be a branch of Physics and Electrical Engineering.

The nonlinear behaviour of active components and their ability to control electron flows makes amplification of weak signals possible, and electronics is widely used in information processing, telecommunication, and signal processing. The ability of electronic devices to act as switches makes digital information processing possible. Interconnection technologies such as circuit boards, electronics packaging technology, and other varied forms of communication infrastructure complete circuit functionality and transform the mixed components into a regular working system.

Electronics is distinct from electrical and electro-mechanical science and technology, which deal with the generation, distribution, switching, storage, and conversion of electrical energy to and from other energy forms using wires, motors, generators, batteries, switches, relays, transformers, resistors, and other passive components. This distinction started around 1906 with the invention by Lee De Forest of the triode, which made electrical amplification of weak radio signals and audio signals possible with a non-mechanical device. Until 1950 this field was called "radio technology" because its principal application was the design and theory of radio transmitters, receivers, and vacuum tubes.

Today, most electronic devices use semiconductor components to perform electron control. The study of semiconductor devices and related technology is considered a branch of solid-state physics, whereas the design and construction of electronic circuits to solve practical problems come under electronics engineering. This article focuses on engineering aspects of electronics.

Branches of Electronics

Electronics has branches as follows:

1. Digital electronics

2. Analogue electronics

3. Microelectronics

4. Circuit design

5. Integrated circuits

6. Optoelectronics

7. Semiconductor devices

8. Embedded systems

Electronic Devices and Components

Electronics Technician performing a voltage check on a power circuit card in the air navigation equipment room aboard the aircraft carrier USS *Abraham Lincoln* (CVN 72).

An electronic component is any physical entity in an electronic system used to affect the electrons or their associated fields in a manner consistent with the intended function of the electronic system. Components are generally intended to be connected together, usually by being soldered to a printed circuit board (PCB), to create an electronic circuit with a particular function (for example an amplifier, radio receiver, or oscillator). Components may be packaged singly, or in more complex groups as integrated circuits. Some common electronic components are capacitors, inductors, resistors, diodes, transistors, etc. Components are often categorized as active (e.g. transistors and thyristors) or passive (e.g. resistors, diodes, inductors and capacitors).

History of Electronic Components

Vacuum tubes (Thermionic valves) were among the earliest electronic components. They were almost solely responsible for the electronics revolution of the first half of the Twentieth Century.

They took electronics from parlor tricks and gave us radio, television, phonographs, radar, long distance telephony and much more. They played a leading role in the field of microwave and high power transmission as well as television receivers until the middle of the 1980s. Since that time, solid state devices have all but completely taken over. Vacuum tubes are still used in some specialist applications such as high power RF amplifiers, cathode ray tubes, specialist audio equipment, guitar amplifiers and some microwave devices.

In April 1955 the IBM 608 was the first IBM product to use transistor circuits without any vacuum tubes and is believed to be the world's first all-transistorized calculator to be manufactured for the commercial market. The 608 contained more than 3,000 germanium transistors. Thomas J. Watson Jr. ordered all future IBM products to use transistors in their design. From that time on transistors were almost exclusively used for computer logic and peripherals.

Types of Circuits

Circuits and components can be divided into two groups: analog and digital. A particular device may consist of circuitry that has one or the other or a mix of the two types.

Analog Circuits

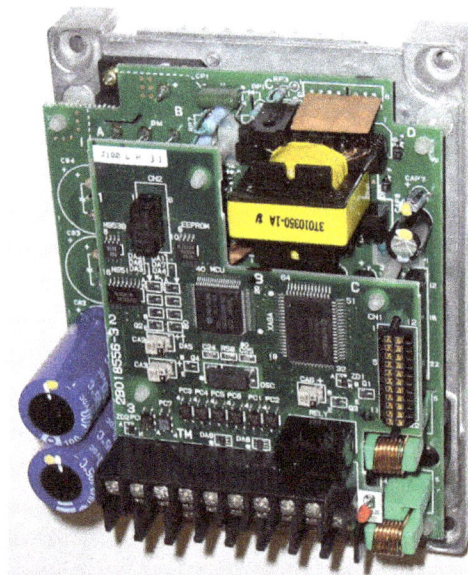

Hitachi J100 adjustable frequency drive chassis

Most analog electronic appliances, such as radio receivers, are constructed from combinations of a few types of basic circuits. Analog circuits use a continuous range of voltage or current as opposed to discrete levels as in digital circuits.

The number of different analog circuits so far devised is huge, especially because a 'circuit' can be defined as anything from a single component, to systems containing thousands of components.

Analog circuits are sometimes called linear circuits although many non-linear effects are used in analog circuits such as mixers, modulators, etc. Good examples of analog circuits include vacuum tube and transistor amplifiers, operational amplifiers and oscillators.

One rarely finds modern circuits that are entirely analog. These days analog circuitry may use digital or even microprocessor techniques to improve performance. This type of circuit is usually called "mixed signal" rather than analog or digital.

Sometimes it may be difficult to differentiate between analog and digital circuits as they have elements of both linear and non-linear operation. An example is the comparator which takes in a continuous range of voltage but only outputs one of two levels as in a digital circuit. Similarly, an overdriven transistor amplifier can take on the characteristics of a controlled switch having essentially two levels of output. In fact, many digital circuits are actually implemented as variations of analog circuits similar to this example—after all, all aspects of the real physical world are essentially analog, so digital effects are only realized by constraining analog behavior.

Digital Circuits

Digital circuits are electric circuits based on a number of discrete voltage levels. Digital circuits are the most common physical representation of Boolean algebra, and are the basis of all digital computers. To most engineers, the terms "digital circuit", "digital system" and "logic" are interchangeable in the context of digital circuits. Most digital circuits use a binary system with two voltage levels labeled "0" and "1". Often logic "0" will be a lower voltage and referred to as "Low" while logic "1" is referred to as "High". However, some systems use the reverse definition ("0" is "High") or are current based. Quite often the logic designer may reverse these definitions from one circuit to the next as he sees fit to facilitate his design. The definition of the levels as "0" or "1" is arbitrary.

Ternary (with three states) logic has been studied, and some prototype computers made.

Computers, electronic clocks, and programmable logic controllers (used to control industrial processes) are constructed of digital circuits. Digital signal processors are another example.

Building blocks:

- Logic gates

- Adders

- Flip-flops

- Counters

- Registers

- Multiplexers

- Schmitt triggers

Highly integrated devices:

- Microprocessors

- Microcontrollers

- Application-specific integrated circuit (ASIC)

- Digital signal processor (DSP)

- Field-programmable gate array (FPGA)

Heat Dissipation and Thermal Management

Heat generated by electronic circuitry must be dissipated to prevent immediate failure and improve long term reliability. Heat dissipation is mostly achieved by passive conduction/convection. Means to achieve greater dissipation include heat sinks and fans for air cooling, and other forms of computer cooling such as water cooling. These techniques use convection, conduction, and radiation of heat energy.

Noise

Electronic noise is defined as unwanted disturbances superposed on a useful signal that tend to obscure its information content. Noise is not the same as signal distortion caused by a circuit. Noise is associated with all electronic circuits. Noise may be electromagnetically or thermally generated, which can be decreased by lowering the operating temperature of the circuit. Other types of noise, such as shot noise cannot be removed as they are due to limitations in physical properties.

Electronics Theory

Mathematical methods are integral to the study of electronics. To become proficient in electronics it is also necessary to become proficient in the mathematics of circuit analysis.

Circuit analysis is the study of methods of solving generally linear systems for unknown variables such as the voltage at a certain node or the current through a certain branch of a network. A common analytical tool for this is the SPICE circuit simulator.

Also important to electronics is the study and understanding of electromagnetic field theory.

Electronics Lab

Due to the complex nature of electronics theory, laboratory experimentation is an important part of the development of electronic devices. These experiments are used to test or verify the engineer's design and detect errors. Historically, electronics labs have consisted of electronics devices and equipment located in a physical space, although in more recent years the trend has been towards electronics lab simulation software, such as CircuitLogix, Multisim, and PSpice.

Computer Aided Design (CAD)

Today's electronics engineers have the ability to design circuits using premanufactured building blocks such as power supplies, semiconductors (i.e. semiconductor devices, such as transistors), and integrated circuits. Electronic design automation software programs include schematic capture programs and printed circuit board design programs. Popular names in the EDA software world are NI Multisim, Cadence (ORCAD), EAGLE PCB and Schematic, Mentor (PADS PCB and LOGIC Schematic), Altium (Protel), LabCentre Electronics (Proteus), gEDA, KiCad and many others.

Construction Methods

Many different methods of connecting components have been used over the years. For instance, early electronics often used point to point wiring with components attached to wooden breadboards to construct circuits. Cordwood construction and wire wrap were other methods used. Most modern day electronics now use printed circuit boards made of materials such as FR4, or the cheaper (and less hard-wearing) Synthetic Resin Bonded Paper (SRBP, also known as Paxoline/Paxolin (trade marks) and FR2) - characterised by its brown colour. Health and environmental concerns associated with electronics assembly have gained increased attention in recent years, especially for products destined to the European Union, with its Restriction of Hazardous Substances Directive (RoHS) and Waste Electrical and Electronic Equipment Directive (WEEE), which went into force in July 2006.

Degradation

Rasberry crazy ants have been known to consume the insides of electrical wiring, and nest inside of electronics; they prefer DC to AC currents. This behavior is not well understood by scientists.

Digital Electronics

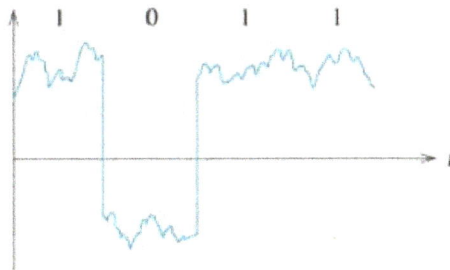

A digital signal has two or more distinguishable waveforms, in this example,
high voltage and low voltages, each of which can be mapped onto a digit.

Digital electronics or digital (electronic) circuits are electronics that handle digital signals (discrete bands of analog levels) rather than by continuous ranges as used in analog electronics. All levels within a band of values represent the same information state. Because of this discretization, relatively small changes to the analog signal levels due to manufacturing tolerance, signal attenuation or noise do not leave the discrete envelope, and as a result are ignored by signal state sensing circuitry.

In most cases, the number of these states is two, and they are represented by two voltage bands: one near a reference value (typically termed as "ground" or zero volts), and the other a value near the supply voltage. These correspond to the *false* and *true* values of the Boolean domain respectively. Digital techniques are useful because it is easier to get an electronic device to switch into one of a number of known states than to accurately reproduce a continuous range of values.

Digital electronic circuits are usually made from large assemblies of logic gates, simple electronic representations of Boolean logic functions.

History

The binary number system was refined by Gottfried Wilhelm Leibniz (published in 1705) and he also established that by using the binary system, the principles of arithmetic and logic could be combined. Digital logic as we know it was the brain-child of George Boole, in the mid 19th century. Boole died young, but his ideas lived on. In an 1886 letter, Charles Sanders Peirce described how logical operations could be carried out by electrical switching circuits. Eventually, vacuum tubes replaced relays for logic operations. Lee De Forest's modification, in 1907, of the Fleming valve can be used as an AND logic gate. Ludwig Wittgenstein introduced a version of the 16-row truth table as proposition 5.101 of *Tractatus Logico-Philosophicus* (1921). Walther Bothe, inventor of the coincidence circuit, got part of the 1954 Nobel Prize in physics, for the first modern electronic AND gate in 1924.

Mechanical analog computers started appearing in the first century and were later used in the medieval era for astronomical calculations. In World War II, mechanical analog computers were used for specialized military applications such as calculating torpedo aiming. During this time the first electronic digital computers were developed. Originally they were the size of a large room, consuming as much power as several hundred modern personal computers (PCs).

The Z3 was an electromechanical computer designed by Konrad Zuse, finished in 1941. It was the world's first working programmable, fully automatic digital computer. Its operation was facilitated by the invention of the vacuum tube in 1904 by John Ambrose Fleming.

Purely electronic circuit elements soon replaced their mechanical and electromechanical equivalents, at the same time that digital calculation replaced analog. The bipolar junction transistor was invented in 1947. From 1955 onwards transistors replaced vacuum tubes in computer designs, giving rise to the "second generation" of computers.

Compared to vacuum tubes, transistors have many advantages: they are smaller, and require less power than vacuum tubes, so give off less heat. Silicon junction transistors were much more reliable than vacuum tubes and had longer, indefinite, service life. Transistorized computers could contain tens of thousands of binary logic circuits in a relatively compact space.

At the University of Manchester, a team under the leadership of Tom Kilburn designed and built a machine using the newly developed transistors instead of valves. Their first transistorised computer and the first in the world, was operational by 1953, and a second version was completed there in April 1955.

While working at Texas Instruments, Jack Kilby recorded his initial ideas concerning the integrated circuit in July 1958, successfully demonstrating the first working integrated example on 12 September 1958. This new technique allowed for quick, low-cost fabrication of complex circuits by having a set of electronic circuits on one small plate ("chip") of semiconductor material, normally silicon.

In the early days of simple integrated circuits, the technology's large scale limited each chip to only a few transistors, and the low degree of integration meant the design process was relatively simple. Manufacturing yields were also quite low by today's standards. As the technology progressed, millions, then billions of transistors could be placed on one chip, and good designs required thorough planning, giving rise to new design methods.

Properties

An advantage of digital circuits when compared to analog circuits is that signals represented digitally can be transmitted without degradation due to noise. For example, a continuous audio signal transmitted as a sequence of 1s and 0s, can be reconstructed without error, provided the noise picked up in transmission is not enough to prevent identification of the 1s and 0s. An hour of music can be stored on a compact disc using about 6 billion binary digits.

In a digital system, a more precise representation of a signal can be obtained by using more binary digits to represent it. While this requires more digital circuits to process the signals, each digit is handled by the same kind of hardware, resulting in an easily scalable system. In an analog system, additional resolution requires fundamental improvements in the linearity and noise characteristics of each step of the signal chain.

Computer-controlled digital systems can be controlled by software, allowing new functions to be added without changing hardware. Often this can be done outside of the factory by updating the product's software. So, the product's design errors can be corrected after the product is in a customer's hands.

Information storage can be easier in digital systems than in analog ones. The noise-immunity of digital systems permits data to be stored and retrieved without degradation. In an analog system, noise from aging and wear degrade the information stored. In a digital system, as long as the total noise is below a certain level, the information can be recovered perfectly.

Even when more significant noise is present, the use of redundancy permits the recovery of the original data provided too many errors do not occur.

In some cases, digital circuits use more energy than analog circuits to accomplish the same tasks, thus producing more heat which increases the complexity of the circuits such as the inclusion of heat sinks. In portable or battery-powered systems this can limit use of digital systems.

For example, battery-powered cellular telephones often use a low-power analog front-end to amplify and tune in the radio signals from the base station. However, a base station has grid power and can use power-hungry, but very flexible software radios. Such base stations can be easily reprogrammed to process the signals used in new cellular standards.

Digital circuits are sometimes more expensive, especially in small quantities.

Most useful digital systems must translate from continuous analog signals to discrete digital signals. This causes quantization errors. Quantization error can be reduced if the system stores enough digital data to represent the signal to the desired degree of fidelity. The Nyquist-Shannon sampling theorem provides an important guideline as to how much digital data is needed to accurately portray a given analog signal.

In some systems, if a single piece of digital data is lost or misinterpreted, the meaning of large blocks of related data can completely change. Because of the cliff effect, it can be difficult for users to tell if a particular system is right on the edge of failure, or if it can tolerate much more noise before failing.

Digital fragility can be reduced by designing a digital system for robustness. For example, a parity bit or other error management method can be inserted into the signal path. These schemes help the system detect errors, and then either correct the errors, or at least ask for a new copy of the data. In a state-machine, the state transition logic can be designed to catch unused states and trigger a reset sequence or other error recovery routine.

Digital memory and transmission systems can use techniques such as error detection and correction to use additional data to correct any errors in transmission and storage.

On the other hand, some techniques used in digital systems make those systems more vulnerable to single-bit errors. These techniques are acceptable when the underlying bits are reliable enough that such errors are highly unlikely.

A single-bit error in audio data stored directly as linear pulse code modulation (such as on a CD-ROM) causes, at worst, a single click. Instead, many people use audio compression to save storage space and download time, even though a single-bit error may corrupt the entire song.

Construction

A binary clock, hand-wired on breadboards

A digital circuit is typically constructed from small electronic circuits called logic gates that can be used to create combinational logic. Each logic gate is designed to perform a function of boolean logic when acting on logic signals. A logic gate is generally created from one or more electrically controlled switches, usually transistors but thermionic valves have seen historic use. The output of a logic gate can, in turn, control or feed into more logic gates.

Integrated circuits consist of multiple transistors on one silicon chip, and are the least expensive way to make large number of interconnected logic gates. Integrated circuits are usually designed by engineers using electronic design automation software to perform some type of function.

Integrated circuits are usually interconnected on a printed circuit board which is a board which holds electrical components, and connects them together with copper traces.

Design

Each logic symbol is represented by a different shape. The actual set of shapes was introduced in

1984 under IEEE/ANSI standard 91-1984. "The logic symbol given under this standard are being increasingly used now and have even started appearing in the literature published by manufacturers of digital integrated circuits."

Another form of digital circuit is constructed from lookup tables, (many sold as "programmable logic devices", though other kinds of PLDs exist). Lookup tables can perform the same functions as machines based on logic gates, but can be easily reprogrammed without changing the wiring. This means that a designer can often repair design errors without changing the arrangement of wires. Therefore, in small volume products, programmable logic devices are often the preferred solution. They are usually designed by engineers using electronic design automation software.

When the volumes are medium to large, and the logic can be slow, or involves complex algorithms or sequences, often a small microcontroller is programmed to make an embedded system. These are usually programmed by software engineers.

When only one digital circuit is needed, and its design is totally customized, as for a factory production line controller, the conventional solution is a programmable logic controller, or PLC. These are usually programmed by electricians, using ladder logic.

Structure of Digital Systems

Engineers use many methods to minimize logic functions, in order to reduce the circuit's complexity. When the complexity is less, the circuit also has fewer errors and less electronics, and is therefore less expensive.

The most widely used simplification is a minimization algorithm like the Espresso heuristic logic minimizer within a CAD system, although historically, binary decision diagrams, an automated Quine–McCluskey algorithm, truth tables, Karnaugh maps, and Boolean algebra have been used.

Representation

Representations are crucial to an engineer's design of digital circuits. Some analysis methods only work with particular representations.

The classical way to represent a digital circuit is with an equivalent set of logic gates. Another way, often with the least electronics, is to construct an equivalent system of electronic switches (usually transistors). One of the easiest ways is to simply have a memory containing a truth table. The inputs are fed into the address of the memory, and the data outputs of the memory become the outputs.

For automated analysis, these representations have digital file formats that can be processed by computer programs. Most digital engineers are very careful to select computer programs ("tools") with compatible file formats.

Combinational vs. Sequential

To choose representations, engineers consider types of digital systems. Most digital systems divide into "combinational systems" and "sequential systems." A combinational system always presents the same output when given the same inputs. It is basically a representation of a set of logic functions, as already discussed.

A sequential system is a combinational system with some of the outputs fed back as inputs. This makes the digital machine perform a "sequence" of operations. The simplest sequential system is probably a flip flop, a mechanism that represents a binary digit or "bit".

Sequential systems are often designed as state machines. In this way, engineers can design a system's gross behavior, and even test it in a simulation, without considering all the details of the logic functions.

Sequential systems divide into two further subcategories. "Synchronous" sequential systems change state all at once, when a "clock" signal changes state. "Asynchronous" sequential systems propagate changes whenever inputs change. Synchronous sequential systems are made of well-characterized asynchronous circuits such as flip-flops, that change only when the clock changes, and which have carefully designed timing margins.

Synchronous Systems

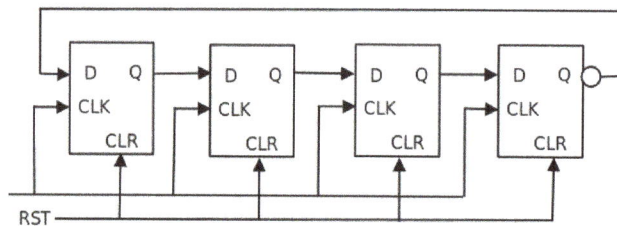

A 4-bit ring counter using D-type flip flops is an example of synchronous logic.
Each device is connected to the clock signal, and update together.

The usual way to implement a synchronous sequential state machine is to divide it into a piece of combinational logic and a set of flip flops called a "state register." Each time a clock signal ticks, the state register captures the feedback generated from the previous state of the combinational logic, and feeds it back as an unchanging input to the combinational part of the state machine. The fastest rate of the clock is set by the most time-consuming logic calculation in the combinational logic.

The state register is just a representation of a binary number. If the states in the state machine are numbered (easy to arrange), the logic function is some combinational logic that produces the number of the next state.

Asynchronous Systems

As of 2014, almost all digital machines are synchronous designs because it is easier to create and verify a synchronous design. However, asynchronous logic is thought can be superior because its speed is not constrained by an arbitrary clock; instead, it runs at the maximum speed of its logic gates. Building an asynchronous system using faster parts makes the circuit faster.

Many systems need circuits that allow external unsynchronized signals to enter synchronous logic circuits. These are inherently asynchronous in their design and must be analyzed as such. Examples of widely used asynchronous circuits include synchronizer flip-flops, switch debouncers and arbiters.

Asynchronous logic components can be hard to design because all possible states, in all possible timings must be considered. The usual method is to construct a table of the minimum and maximum time that each such state can exist, and then adjust the circuit to minimize the number of such states. Then the designer must force the circuit to periodically wait for all of its parts to enter a compatible state (this is called "self-resynchronization"). Without such careful design, it is easy to accidentally produce asynchronous logic that is "unstable," that is, real electronics will have unpredictable results because of the cumulative delays caused by small variations in the values of the electronic components.

Register Transfer Systems

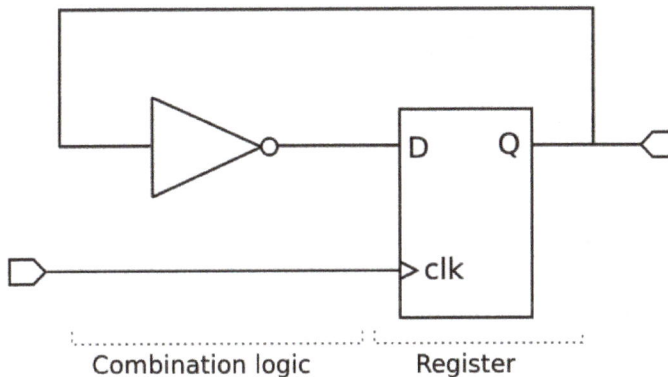

Example of a simple circuit with a toggling output. The inverter forms the combinational logic in this circuit, and the register holds the state.

Many digital systems are data flow machines. These are usually designed using synchronous register transfer logic, using hardware description languages such as VHDL or Verilog.

In register transfer logic, binary numbers are stored in groups of flip flops called registers. The outputs of each register are a bundle of wires called a "bus" that carries that number to other calculations. A calculation is simply a piece of combinational logic. Each calculation also has an output bus, and these may be connected to the inputs of several registers. Sometimes a register will have a multiplexer on its input, so that it can store a number from any one of several buses. Alternatively, the outputs of several items may be connected to a bus through buffers that can turn off the output of all of the devices except one. A sequential state machine controls when each register accepts new data from its input.

Asynchronous register-transfer systems (such as computers) have a general solution. In the 1980s, some researchers discovered that almost all synchronous register-transfer machines could be converted to asynchronous designs by using first-in-first-out synchronization logic. In this scheme, the digital machine is characterized as a set of data flows. In each step of the flow, an asynchronous "synchronization circuit" determines when the outputs of that step are valid, and presents a signal that says, "grab the data" to the stages that use that stage's inputs. It turns out that just a few relatively simple synchronization circuits are needed.

Computer Design

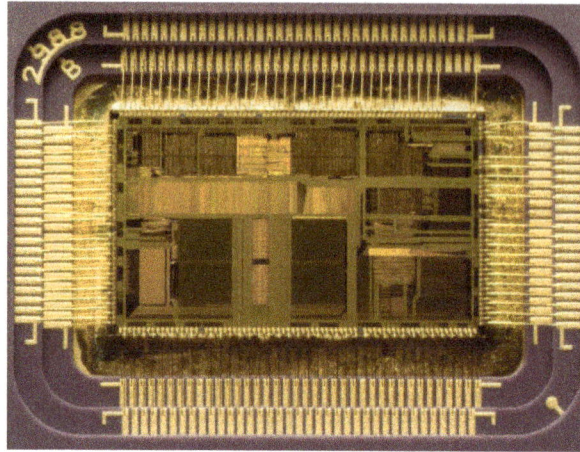

Intel 80486DX2 microprocessor

The most general-purpose register-transfer logic machine is a computer. This is basically an automatic binary abacus. The control unit of a computer is usually designed as a microprogram run by a microsequencer. A microprogram is much like a player-piano roll. Each table entry or "word" of the microprogram commands the state of every bit that controls the computer. The sequencer then counts, and the count addresses the memory or combinational logic machine that contains the microprogram. The bits from the microprogram control the arithmetic logic unit, memory and other parts of the computer, including the microsequencer itself. A "specialized computer" is usually a conventional computer with special-purpose control logic or microprogram.

In this way, the complex task of designing the controls of a computer is reduced to a simpler task of programming a collection of much simpler logic machines.

Almost all computers are synchronous. However, true asynchronous computers have also been designed. One example is the Aspida DLX core. Another was offered by ARM Holdings. Speed advantages have not materialized, because modern computer designs already run at the speed of their slowest component, usually memory. These do use somewhat less power because a clock distribution network is not needed. An unexpected advantage is that asynchronous computers do not produce spectrally-pure radio noise, so they are used in some mobile-phone base-station controllers. They may be more secure in cryptographic applications because their electrical and radio emissions can be more difficult to decode.

Computer Architecture

Computer architecture is a specialized engineering activity that tries to arrange the registers, calculation logic, buses and other parts of the computer in the best way for some purpose. Computer architects have applied large amounts of ingenuity to computer design to reduce the cost and increase the speed and immunity to programming errors of computers. An increasingly common goal is to reduce the power used in a battery-powered computer system, such as a cell-phone. Many computer architects serve an extended apprenticeship as microprogrammers.

Design Issues in Digital Circuits

Digital circuits are made from analog components. The design must assure that the analog nature of the components doesn't dominate the desired digital behavior. Digital systems must manage noise and timing margins, parasitic inductances and capacitances, and filter power connections.

Bad designs have intermittent problems such as "glitches", vanishingly fast pulses that may trigger some logic but not others, "runt pulses" that do not reach valid "threshold" voltages, or unexpected ("undecoded") combinations of logic states.

Additionally, where clocked digital systems interface to analog systems or systems that are driven from a different clock, the digital system can be subject to metastability where a change to the input violates the set-up time for a digital input latch. This situation will self-resolve, but will take a random time, and while it persists can result in invalid signals being propagated within the digital system for a short time.

Since digital circuits are made from analog components, digital circuits calculate more slowly than low-precision analog circuits that use a similar amount of space and power. However, the digital circuit will calculate more repeatably, because of its high noise immunity. On the other hand, in the high-precision domain (for example, where 14 or more bits of precision are needed), analog circuits require much more power and area than digital equivalents.

Automated Design Tools

To save costly engineering effort, much of the effort of designing large logic machines has been automated. The computer programs are called "electronic design automation tools" or just "EDA."

Simple truth table-style descriptions of logic are often optimized with EDA that automatically produces reduced systems of logic gates or smaller lookup tables that still produce the desired outputs. The most common example of this kind of software is the Espresso heuristic logic minimizer.

Most practical algorithms for optimizing large logic systems use algebraic manipulations or binary decision diagrams, and there are promising experiments with genetic algorithms and annealing optimizations.

To automate costly engineering processes, some EDA can take state tables that describe state machines and automatically produce a truth table or a function table for the combinational logic of a state machine. The state table is a piece of text that lists each state, together with the conditions controlling the transitions between them and the belonging output signals.

It is common for the function tables of such computer-generated state-machines to be optimized with logic-minimization software such as Minilog.

Often, real logic systems are designed as a series of sub-projects, which are combined using a "tool flow." The tool flow is usually a "script," a simplified computer language that can invoke the software design tools in the right order.

Tool flows for large logic systems such as microprocessors can be thousands of commands long, and combine the work of hundreds of engineers.

Writing and debugging tool flows is an established engineering specialty in companies that produce digital designs. The tool flow usually terminates in a detailed computer file or set of files that describe how to physically construct the logic. Often it consists of instructions to draw the transistors and wires on an integrated circuit or a printed circuit board.

Parts of tool flows are "debugged" by verifying the outputs of simulated logic against expected inputs. The test tools take computer files with sets of inputs and outputs, and highlight discrepancies between the simulated behavior and the expected behavior.

Once the input data is believed correct, the design itself must still be verified for correctness. Some tool flows verify designs by first producing a design, and then scanning the design to produce compatible input data for the tool flow. If the scanned data matches the input data, then the tool flow has probably not introduced errors.

The functional verification data are usually called "test vectors". The functional test vectors may be preserved and used in the factory to test that newly constructed logic works correctly. However, functional test patterns don't discover common fabrication faults. Production tests are often designed by software tools called "test pattern generators". These generate test vectors by examining the structure of the logic and systematically generating tests for particular faults. This way the fault coverage can closely approach 100%, provided the design is properly made testable.

Once a design exists, and is verified and testable, it often needs to be processed to be manufacturable as well. Modern integrated circuits have features smaller than the wavelength of the light used to expose the photoresist. Manufacturability software adds interference patterns to the exposure masks to eliminate open-circuits, and enhance the masks' contrast.

Design for Testability

There are several reasons for testing a logic circuit. When the circuit is first developed, it is necessary to verify that the design circuit meets the required functional and timing specifications. When multiple copies of a correctly designed circuit are being manufactured, it is essential to test each copy to ensure that the manufacturing process has not introduced any flaws.

A large logic machine (say, with more than a hundred logical variables) can have an astronomical number of possible states. Obviously, in the factory, testing every state is impractical if testing each state takes a microsecond, and there are more states than the number of microseconds since the universe began. Unfortunately, this ridiculous-sounding case is typical.

Fortunately, large logic machines are almost always designed as assemblies of smaller logic machines. To save time, the smaller sub-machines are isolated by permanently installed "design for test" circuitry, and are tested independently.

One common test scheme known as "scan design" moves test bits serially (one after another) from external test equipment through one or more serial shift registers known as "scan chains". Serial scans have only one or two wires to carry the data, and minimize the physical size and expense of the infrequently used test logic.

After all the test data bits are in place, the design is reconfigured to be in "normal mode" and one or more clock pulses are applied, to test for faults (e.g. stuck-at low or stuck-at high) and capture the test result into flip-flops and/or latches in the scan shift register(s). Finally, the result of the test is shifted out to the block boundary and compared against the predicted "good machine" result.

In a board-test environment, serial to parallel testing has been formalized with a standard called "JTAG" (named after the "Joint Test Action Group" that proposed it).

Another common testing scheme provides a test mode that forces some part of the logic machine to enter a "test cycle." The test cycle usually exercises large independent parts of the machine.

Trade-offs

Several numbers determine the practicality of a system of digital logic: cost, reliability, fanout and speed. Engineers explored numerous electronic devices to get an ideal combination of these traits.

Cost

The cost of a logic gate is crucial, primarily because very many gates are needed to build a computer or other advanced digital system and because the more gates can be used, the more capable and/or fast the machine can be. Since the majority of a digital computer is simply an interconnected network of logic gates, the overall cost of building a computer correlates strongly with the price per logic gate. In the 1930s, the earliest digital logic systems were constructed from telephone relays because these were inexpensive and relatively reliable. After that, engineers always used the cheapest available electronic switches that could still fulfill the requirements.

The earliest integrated circuits were a happy accident. They were constructed not to save money, but to save weight, and permit the Apollo Guidance Computer to control an inertial guidance system for a spacecraft. The first integrated circuit logic gates cost nearly $50 (in 1960 dollars, when an engineer earned $10,000/year). To everyone's surprise, by the time the circuits were mass-produced, they had become the least-expensive method of constructing digital logic. Improvements in this technology have driven all subsequent improvements in cost.

With the rise of integrated circuits, reducing the absolute number of chips used represented another way to save costs. The goal of a designer is not just to make the simplest circuit, but to keep the component count down. Sometimes this results in more complicated designs with respect to the underlying digital logic but nevertheless reduces the number of components, board size, and even power consumption. A major motive for reducing component count on printed circuit boards is to reduce the manufacturing defect rate and increase reliability, as every soldered connection is a potentially bad one, so the defect and failure rates tend to increase along with the total number of component pins.

For example, in some logic families, NAND gates are the simplest digital gate to build. All other logical operations can be implemented by NAND gates. If a circuit already required a single NAND gate, and a single chip normally carried four NAND gates, then the remaining gates could be used to implement other logical operations like logical and. This could eliminate the need for a separate chip containing those different types of gates.

Reliability

The "reliability" of a logic gate describes its mean time between failure (MTBF). Digital machines often have millions of logic gates. Also, most digital machines are "optimized" to reduce their cost. The result is that often, the failure of a single logic gate will cause a digital machine to stop working. It is possible to design machines to be more reliable by using redundant logic which will not malfunction as a result of the failure of any single gate (or even any two, three, or four gates), but this necessarily entails using more components, which raises the financial cost and also usually increases the weight of the machine and may increase the power it consumes.

Digital machines first became useful when the MTBF for a switch got above a few hundred hours. Even so, many of these machines had complex, well-rehearsed repair procedures, and would be nonfunctional for hours because a tube burned-out, or a moth got stuck in a relay. Modern transistorized integrated circuit logic gates have MTBFs greater than 82 billion hours ($8.2 \cdot 10^{10}$ hours), and need them because they have so many logic gates.

Fanout

Fanout describes how many logic inputs can be controlled by a single logic output without exceeding the electrical current ratings of the gate outputs. The minimum practical fanout is about five. Modern electronic logic gates using CMOS transistors for switches have fanouts near fifty, and can sometimes go much higher.

Speed

The "switching speed" describes how many times per second an inverter (an electronic representation of a "logical not" function) can change from true to false and back. Faster logic can accomplish more operations in less time. Digital logic first became useful when switching speeds got above 50 Hz, because that was faster than a team of humans operating mechanical calculators. Modern electronic digital logic routinely switches at 5 GHz ($5 \cdot 10^9$ Hz), and some laboratory systems switch at more than 1 THz ($1 \cdot 10^{12}$ Hz).

Logic Families

Design started with relays. Relay logic was relatively inexpensive and reliable, but slow. Occasionally a mechanical failure would occur. Fanouts were typically about 10, limited by the resistance of the coils and arcing on the contacts from high voltages.

Later, vacuum tubes were used. These were very fast, but generated heat, and were unreliable because the filaments would burn out. Fanouts were typically 5...7, limited by the heating from the tubes' current. In the 1950s, special "computer tubes" were developed with filaments that omitted volatile elements like silicon. These ran for hundreds of thousands of hours.

The first semiconductor logic family was resistor–transistor logic. This was a thousand times more reliable than tubes, ran cooler, and used less power, but had a very low fan-in of 3. Diode–transistor logic improved the fanout up to about 7, and reduced the power. Some DTL designs used two power-supplies with alternating layers of NPN and PNP transistors to increase the fanout.

Transistor–transistor logic (TTL) was a great improvement over these. In early devices, fanout improved to 10, and later variations reliably achieved 20. TTL was also fast, with some variations achieving switching times as low as 20 ns. TTL is still used in some designs.

Emitter coupled logic is very fast but uses a lot of power. It was extensively used for high-performance computers made up of many medium-scale components (such as the Illiac IV).

By far, the most common digital integrated circuits built today use CMOS logic, which is fast, offers high circuit density and low-power per gate. This is used even in large, fast computers, such as the IBM System z.

Recent Developments

In 2009, researchers discovered that memristors can implement a boolean state storage (similar to a flip flop, implication and logical inversion), providing a complete logic family with very small amounts of space and power, using familiar CMOS semiconductor processes.

The discovery of superconductivity has enabled the development of rapid single flux quantum (RSFQ) circuit technology, which uses Josephson junctions instead of transistors. Most recently, attempts are being made to construct purely optical computing systems capable of processing digital information using nonlinear optical elements.

Analogue Electronics

Analogue electronics (also spelled analog electronics) are electronic systems with a continuously variable signal, in contrast to digital electronics where signals usually take only two levels. The term "analogue" describes the proportional relationship between a signal and a voltage or current that represents the signal.

Analogue Signals

An analogue signal uses some attribute of the medium to convey the signal's information. For example, an aneroid barometer uses the angular position of a needle as the signal to convey the information of changes in atmospheric pressure. Electrical signals may represent information by changing their voltage, current, frequency, or total charge. Information is converted from some other physical form (such as sound, light, temperature, pressure, position) to an electrical signal by a transducer which converts one type of energy into another (e.g. a microphone).

The signals take any value from a given range, and each unique signal value represents different information. Any change in the signal is meaningful, and each level of the signal represents a different level of the phenomenon that it represents. For example, suppose the signal is being used to represent temperature, with one volt representing one degree Celsius. In such a system 10 volts would represent 10 degrees, and 10.1 volts would represent 10.1 degrees.

Another method of conveying an analogue signal is to use modulation. In this, some base carrier signal has one of its properties altered: amplitude modulation (AM) involves altering the amplitude of a sinusoidal voltage waveform by the source information, frequency modulation (FM) changes the frequency. Other techniques, such as phase modulation or changing the phase of the carrier signal, are also used.

In an analogue sound recording, the variation in pressure of a sound striking a microphone creates a corresponding variation in the current passing through it or voltage across it. An increase in the volume of the sound causes the fluctuation of the current or voltage to increase proportionally while keeping the same waveform or shape.

Mechanical, pneumatic, hydraulic and other systems may also use analogue signals.

Inherent Noise

Analogue systems invariably include noise that is random disturbances or variations, some caused by the random thermal vibrations of atomic particles. Since all variations of an analogue signal are significant, any disturbance is equivalent to a change in the original signal and so appears as noise. As the signal is copied and re-copied, or transmitted over long distances, these random variations become more significant and lead to signal degradation. Other sources of noise may include crosstalk from other signals or poorly designed components. These disturbances are reduced by shielding and by using low-noise amplifiers (LNA).

Analogue vs Digital Electronics

Since the information is encoded differently in analogue and digital electronics, the way they process a signal is consequently different. All operations that can be performed on an analogue signal such as amplification, filtering, limiting, and others, can also be duplicated in the digital domain. Every digital circuit is also an analogue circuit, in that the behaviour of any digital circuit can be explained using the rules of analogue circuits.

The first electronic devices invented and mass-produced were analogue. The use of microelectronics has made digital devices cheap and widely available.

Noise

Because of the way information is encoded in analogue circuits, they are much more susceptible to noise than digital circuits, since a small change in the signal can represent a significant change in the information present in the signal and can cause the information present to be lost. Since digital signals take on one of only two different values, a disturbance would have to be about one-half the magnitude of the digital signal to cause an error. This property of digital circuits can be exploited to make signal processing noise-resistant. In digital electronics, because the information is quantized, as long as the signal stays inside a range of values, it represents the same information. Digital circuits use this principle to regenerate the signal at each logic gate, lessening or removing noise.

Precision

A number of factors affect how precise a signal is, mainly the noise present in the original signal and the noise added by processing. Fundamental physical limits such as the shot noise in components limits the resolution of analogue signals. In digital electronics additional precision is obtained by using additional digits to represent the signal. The practical limit in the number of digits is determined by the performance of the analogue-to-digital converter (ADC), since digital operations can usually be performed without loss of precision. The ADC takes an analogue signal and changes it into a series of binary numbers. The ADC may be used in simple digital display devices, e. g., thermometers or light meters but it may also be used in digital sound recording and in data acquisition. However, a digital-to-analogue converter (DAC) is used to change a digital signal to an analogue signal. A DAC takes a series of binary numbers and converts it to an analogue signal. It is common to find a DAC in the gain-control system of an op-amp which in turn may be used to control digital amplifiers and filters.

Design Difficulty

Analogue circuits are typically harder to design, requiring more skill than comparable digital systems. This is one of the main reasons why digital systems have become more common than analogue devices. An analogue circuit was usually designed by hand, and the process is much less automated than for digital systems (at least, until the invention of the Arco compiler in 2016). However, if a digital electronic device is to interact with the real world, it will always need an analogue interface. For example, every digital radio receiver has an analogue preamplifier as the first stage in the receive chain.

Electronic Circuit Design

Electronic circuit design comprises the analysis and synthesis of electronic circuits.

Methods

To design any electrical circuit, either analog or digital, electrical engineers need to be able to predict the voltages and currents at all places within the circuit. Linear circuits, that is, circuits wherein the outputs are linearly dependent on the inputs, can be analyzed by hand using complex analysis. Simple nonlinear circuits can also be analyzed in this way. Specialized software has been created to analyze circuits that are either too complicated or too nonlinear to analyze by hand.

Circuit simulation software allows engineers to design circuits more efficiently, reducing the time cost and risk of error involved in building circuit prototypes. Some of these make use of hardware description languages such as VHDL or Verilog.

Network Simulation Software

More complex circuits are analyzed with circuit simulation software such as SPICE and EMTP.

Linearization Around Operating Point

When faced with a new circuit, the software first tries to find a steady state solution wherein all the nodes conform to Kirchhoff's Current Law *and* the voltages across and through each element of the circuit conform to the voltage/current equations governing that element.

Once the steady state solution is found, the software can analyze the response to perturbations using piecewise approximation, harmonic balance or other methods.

Piece-wise Linear Approximation

Software such as the PLECS interface to Simulink uses piecewise linear approximation of the equations governing the elements of a circuit. The circuit is treated as a completely linear network of ideal diodes. Every time a diode switches from on to off or vice versa, the configuration of the linear network changes. Adding more detail to the approximation of equations increases the accuracy of the simulation, but also increases its running time.

Synthesis

Simple circuits may be designed by connecting a number of elements or functional blocks such as integrated circuits.

More complex digital circuits are typically designed with the aid of computer software. Logic circuits (and sometimes mixed mode circuits) are often described in such hardware description languages as HDL, VHDL or Verilog, then synthesized using a logic synthesis engine.

Π pad

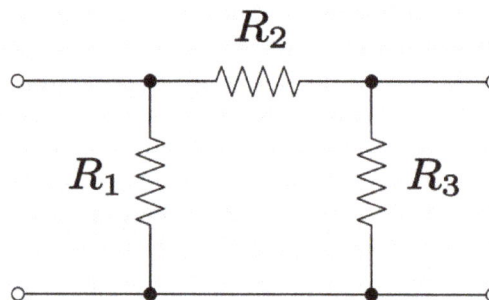

Figure 1. Schematic circuit of a Π-pad attenuator.

The Π pad (pi pad) is a specific type of attenuator circuit in electronics whereby the topology of the circuit is formed in the shape of the Greek letter "Π".

Attenuators are used in electronics to reduce the level of a signal. They are also referred to as pads due to their effect of padding down a signal by analogy with acoustics. Attenuators have a flat frequency response attenuating all frequencies equally in the band they are intended to operate. The attenuator has the opposite task of an amplifier. The topology of an attenuator circuit will usually follow one of the simple filter sections. However, there is no need for more complex circuitry, as there is with filters, due to the simplicity of the frequency response required.

Circuits are required to be balanced or unbalanced depending on the geometry of the transmission lines they are to be used with. For radio frequency applications, the format is often unbalanced, such as coaxial. For audio and telecommunications, balanced circuits are usually required, such as with the twisted pair format. The Π pad is intrinsically an unbalanced circuit. However, it can be converted to a balanced circuit by placing half the series resistance in the return path. Such a circuit is called a box section because the circuit is formed in the shape of a box.

Terminology

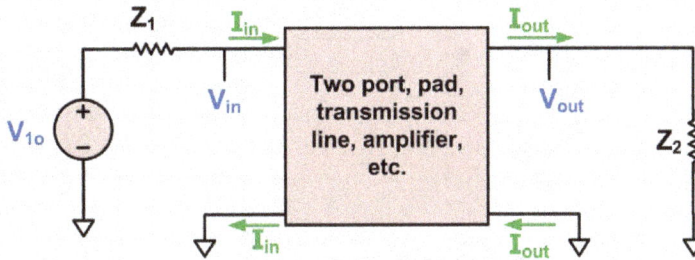

Unbalanced source and load. V1o is open circuit voltage of the source.

An attenuator is a form of a two-port network with a generator connected to one port and a load connected to the other. In all of the circuits given below it is assumed that the generator and load impedances are purely resistive (though not necessarily equal) and that the attenuator circuit is required to perfectly match to these. The symbols used for these impedances are;

Z_1 the impedance of the generator

Z_2 the impedance of the load

Popular values of impedance are 600Ω in telecommucations and audio, 75Ω for video and dipole antennae, 50Ω for RF

The voltage transfer function, A, is,

$$A = \frac{V_{out}}{V_{in}}$$

While the inverse of this is the loss, L, of the attenuator,

$$L = \frac{V_{in}}{V_{out}}$$

The value of attenuation is normally marked on the attenuator as its loss, L_{dB}, in decibels (dB). The relationship with L is;

$$L_{dB} = 20 \log L$$

Popular values of attenuator are 3dB, 6dB, 10dB, 20dB and 40dB.

However, it is often more convenient to express the loss in nepers,

$$L = e^{\gamma}$$

where γ is the attenuation in nepers (one neper is approximately 8.7 dB).

Impedance and Loss

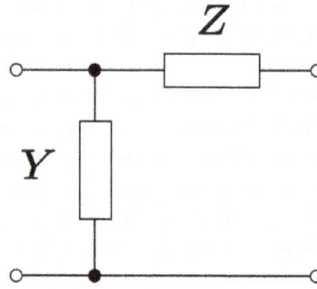

Figure 2. A general L-section circuit with shunt admittance Y and series impedance Z.

The values of resistance of the attenuator's elements can be calculated using image parameter theory. The starting point here is the image impedances of the L section in figure 2. The image admittance of the input is,

$$Y_{i\Pi} = \sqrt{Y^2 + \frac{Y}{Z}}$$

and the image impedance of the output is,

$$Z_{iT} = \sqrt{Z^2 + \frac{Z}{Y}}$$

The loss of the L section when terminated in its image impedances is,

$$L_{L1} = \sqrt{Z_{i\Pi} Y_{iT}} \; e^{\gamma_L}$$

where the image parameter transmission function, γ_L is given by,

$$\gamma_L = \sinh^{-1} \sqrt{ZY}$$

The loss of this L section in the reverse direction is given by,

$$L_{L2} = \sqrt{Z_{iT} Y_{i\Pi}} \; e^{\gamma_L}$$

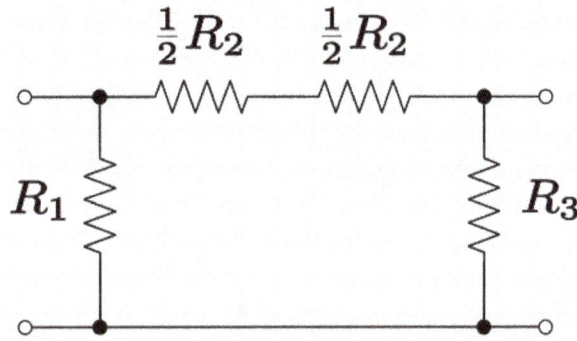

Figure 3. A Π-pad attenuator formed from two symmetrical L sections. Because of the symmetry, $R_1 = R_3$ in this case.

For an attenuator, Z and Y are simple resistors and γ becomes the image parameter attenuation (that is, the attenuation when terminated with the image impedances) in nepers. A Π pad can be viewed as being two L sections back-to-back as shown in figure 3. Most commonly, the generator and load impedances are equal so that $Z_1 = Z_2 = Z_0$ and a symmetrical Π pad is used. In this case, the impedance matching terms inside the square roots all cancel and,

$$L_\Pi = L_{L1}L_{L2} = e^{2\gamma_L} = e^{\gamma_\Pi}$$

Substituting Z and Y for the corresponding resistors,

$$\gamma_\Pi = 2\gamma_L = 2\sinh^{-1}\sqrt{\frac{R_2}{2R_1}}$$

$$\frac{1}{Z_0} = \sqrt{\frac{1}{R_1^2} + \frac{2}{R_1 R_2}}$$

These equations can easily be extended to non-symmetrical cases.

Resistor Values

The equations above find the impedance and loss for an attenuator with given resistor values. The usual requirement in a design is the other way around – the resistor values for a given impedance and loss are needed. These can be found by transposing and substituting the last two equations above;

$$\text{If } Z_0 = Z_1 = Z_2$$

$$R_1 = Z_0 \coth\left(\frac{\gamma_\Pi}{2}\right) = Z_0 \frac{1+A}{1-A} \text{ with } A = \frac{V_{out}}{V_{in}}$$

$$R_2 = \frac{2R_1}{\left(\dfrac{R_1}{Z_0}\right)^2 - 1}$$

O pad

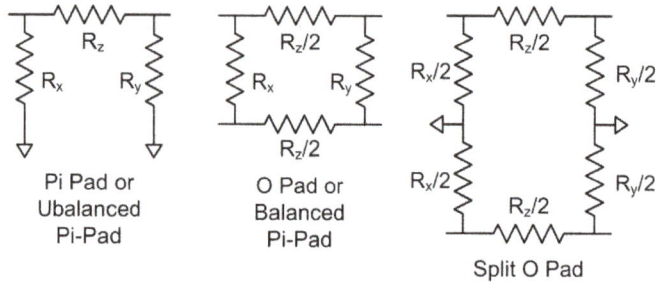

Pi pads, O pads and split O pads

The unbalanced pi pad can be converted to a balanced O pad by putting one half of Rz in each side of a balanced line.

The simple four element O pad attenuates the differential mode signal but does little to attenuate any common mode signal. To insure attenuation of the common mode signal also, a split O pad can be created by splitting and grounding Rx and Ry.

Conversion of Two-port to pi pad

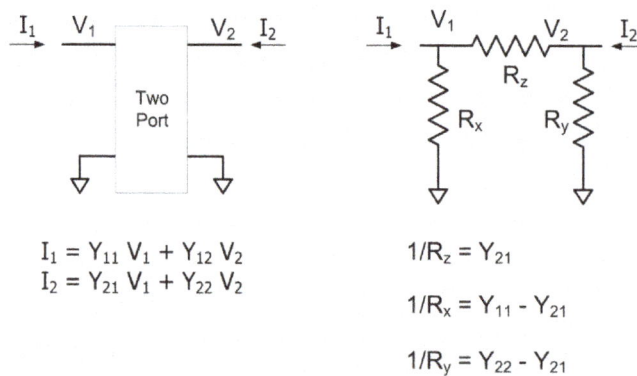

$$I_1 = Y_{11} V_1 + Y_{12} V_2$$
$$I_2 = Y_{21} V_1 + Y_{22} V_2$$

$$1/R_z = Y_{21}$$

$$1/R_x = Y_{11} - Y_{21}$$

$$1/R_y = Y_{22} - Y_{21}$$

Conversion of two-port admittance parameters to pi pad

If a passive two-port can be expressed with admittance parameters, then that two-port is equivalent to a pi pad. In general, the admittance parameters are frequency dependent and not necessarily resistive. In that case the elements of the pi pad would not be simple components. However, in the case where the two-port is purely resistive or substantially resistive over the frequency range of interest, then the two-port can be replaced with a pi pad made of resistors.

Conversion of Tee Pad to pi Pad

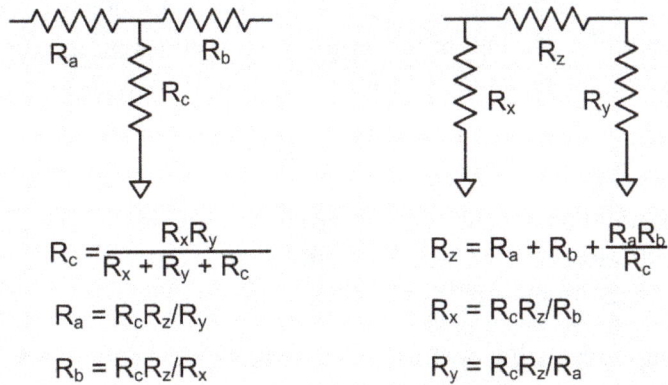

$$R_c = \frac{R_x R_y}{R_x + R_y + R_c}$$

$$R_a = R_c R_z / R_y$$

$$R_b = R_c R_z / R_x$$

$$R_z = R_a + R_b + \frac{R_a R_b}{R_c}$$

$$R_x = R_c R_z / R_b$$

$$R_y = R_c R_z / R_a$$

Conversion of tee pad to pi pad

Pi pads and tee pads are easily converted back and forth.

If one of the pads is composed of only resistors then the other is also composed entirely of resistors.

Electronic Component

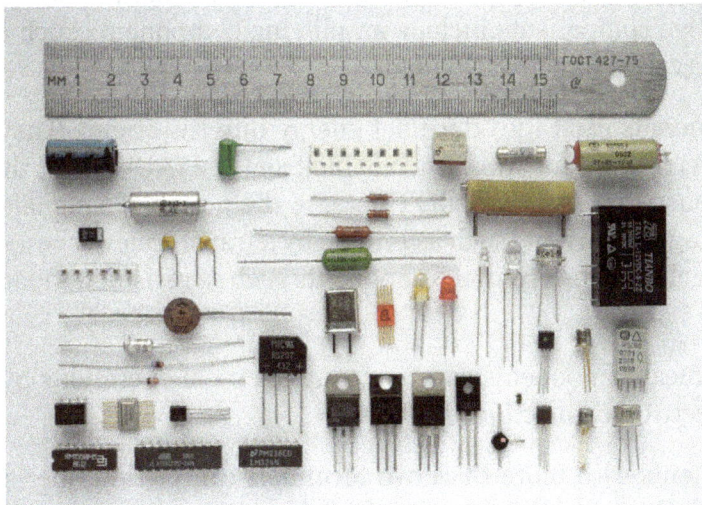

Various electronic components. The ruler at the top is for size comparison.

An electronic component is any basic discrete device or physical entity in an electronic system used to affect electrons or their associated fields. Electronic components are mostly industrial products, available in a singular form and are not to be confused with electrical elements, which are conceptual abstractions representing idealized electronic components.

Electronic components have two or more electrical terminals (or *leads*) aside from antennas which may only have one terminal. These leads connect to create an electronic circuit with a particular function (for example an amplifier, radio receiver, or oscillator). Basic electronic components may be packaged discretely, as arrays or networks of like components, or integrated inside of packages such as semiconductor integrated circuits, hybrid integrated circuits, or thick film devices. The following list of electronic components focuses on the discrete version of these components, treating such packages as components in their own right.

Classification

Components can be classified as passive, active, or electromechanic. The strict physics definition treats passive components as ones that cannot supply energy themselves, whereas a battery would be seen as an active component since it truly acts as a source of energy.

However, electronic engineers who perform circuit analysis use a more restrictive definition of passivity. When only concerned with the energy of signals, it is convenient to ignore the so-called DC circuit and pretend that the power supplying components such as transistors or integrated circuits is absent (as if each such component had its own battery built in), though it may in reality be supplied by the DC circuit. Then, the analysis only concerns the AC circuit, an abstraction that ignores DC voltages and currents (and the power associated with them) present in the real-life circuit. This fiction, for instance, lets us view an oscillator as "producing energy" even though in reality the oscillator consumes even more energy from a DC power supply, which we have chosen to ignore. Under that restriction, we define the terms as used in circuit analysis as:

- Active components rely on a source of energy (usually from the DC circuit, which we have chosen to ignore) and usually can inject power into a circuit, though this is not part of the definition. Active components include amplifying components such as transistors, triode vacuum tubes (valves), and tunnel diodes.

- Passive components can't introduce net energy into the circuit. They also can't rely on a source of power, except for what is available from the (AC) circuit they are connected to. As a consequence they can't amplify (increase the power of a signal), although they may increase a voltage or current (such as is done by a transformer or resonant circuit). Passive components include two-terminal components such as resistors, capacitors, inductors, and transformers.

- Electromechanical components can carry out electrical operations by using moving parts or by using electrical connections

Most passive components with more than two terminals can be described in terms of two-port parameters that satisfy the principle of reciprocity—though there are rare exceptions. In contrast, active components (with more than two terminals) generally lack that property.

Active Components

The electronic components which are perform on the core electronic function like amplification, electronic switching, electronic controlling and the components acts upon a source of current are active component.

Semiconductors

Diodes

Various examples of Light-emitting diodes

Conduct electricity easily in one direction, among more specific behaviors.

- Diode, Rectifier, Bridge rectifier

- Schottky diode, hot carrier diode – super fast diode with lower forward voltage drop

- Zener diode – Passes current in reverse direction to provide a constant voltage reference

- Transient voltage suppression diode (TVS), Unipolar or Bipolar – used to absorb high-voltage spikes

- Varactor, Tuning diode, Varicap, Variable capacitance diode – A diode whose AC capacitance varies according to the DC voltage applied.

- Light-emitting diode (LED) – A diode that emits light

- Photodiode – Passes current in proportion to incident light

    o   Avalanche photodiode Photodiode with internal gain

    o   Solar Cell, photovoltaic cell, PV array or panel, produces power from light

- DIAC (Diode for Alternating Current), Trigger Diode, SIDAC) – Often used to trigger an SCR

- Constant-current diode

- Peltier cooler – A semiconductor heat pump

- Tunnel diode - very fast diode based on quantum mechanical tunneling

Transistors

Transistors were considered the invention of the twentieth century that changed electronic circuits forever. A transistor is a semiconductor device used to amplify and switch electronic signals and electrical power.

- Transistors

    o   Bipolar junction transistor (BJT, or simply "transistor") – NPN or PNP

         □ Photo transistor – Amplified photodetector

- o Darlington transistor – NPN or PNP

         □ Photo Darlington – Amplified photodetector

- o Sziklai pair (Compound transistor, complementary Darlington)
- Field-effect transistor (FET)
  - o JFET (Junction Field-Effect Transistor) – N-CHANNEL or P-CHANNEL
  - o MOSFET (Metal Oxide Semiconductor FET) – N-CHANNEL or P-CHANNEL
  - o MESFET (MEtal Semiconductor FET)
  - o HEMT (High electron mobility transistor)
- Thyristors
  - o Silicon-controlled rectifier (SCR) – Passes current only after triggered by a sufficient control voltage on its gate
  - o TRIAC (TRIode for Alternating Current) – Bidirectional SCR
  - o Unijunction transistor (UJT)
  - o Programmable Unijunction transistor (PUT)
  - o SIT (Static induction transistor)
  - o SITh (Static induction thyristor)
- Composite transistors
  - o IGBT (Insulated-gate bipolar transistor)

Integrated Circuits

- Digital
- Analog
  - o Hall effect sensor –senses a magnetic field
  - o Current sensor – Senses a current through it

Optoelectronic Devices

- Opto-electronics
  - o Opto-Isolator, Opto-Coupler, Photo-Coupler – Photodiode, BJT, JFET, SCR, TRIAC, Zero-crossing TRIAC, Open collector IC, CMOS IC, Solid state relay (SSR)

- o   Opto switch, Opto interrupter, Optical switch, Optical interrupter, Photo switch, Photo interrupter

- o   LED display – Seven-segment display, Sixteen-segment display, Dot-matrix display

Display Technologies

Current:

- Filament lamp (indicator lamp)

- Vacuum fluorescent display (VFD) (preformed characters, 7 segment, starburst)

- Cathode ray tube (CRT) (dot matrix scan, radial scan (e.g. radar), arbitrary scan (e.g. oscilloscope)) (monochrome & colour)

- LCD (preformed characters, dot matrix) (passive, TFT) (monochrome, colour)

- Neon (individual, 7 segment display)

- LED (individual, 7 segment display, starburst display, dot matrix)

- Flap indicator (numeric, preprinted messages)

- Plasma display (dot matrix)

Obsolete:

- Incandescent filament 7 segment display (aka 'Numitron')

- Nixie Tube

- Dekatron (aka glow transfer tube)

- Magic eye tube indicator

- Penetron (a 2 colour see-through CRT)

Vacuum Tubes (Valves)

A vacuum tube is based on current conduction through a vacuum.

- Diode or rectifier tube

Amplifying tubes

- Triode

- Tetrode

- Pentode

- Hexode

- Pentagrid

- Octode

- Microwave tubes

  o Klystron

  o Magnetron

  o Traveling-wave tube

Optical detectors or emitters

- Phototube or Photodiode – tube equivalent of semiconductor photodiode

- Photomultiplier tube – Phototube with internal gain

- Cathode ray tube (CRT) or television picture tube

- Vacuum fluorescent display (VFD) – Modern non-raster sort of small CRT display

- Magic eye tube – Small CRT display used as a tuning meter (obsolete)

- X-ray tube – Produces x-rays

Discharge Devices

- Gas discharge tube

Obsolete:

- Mercury arc rectifier

- Voltage regulator tube

- Nixie tube

- Thyratron

- Ignitron

Power Sources

Sources of electrical power:

- Battery – acid- or alkali-based power supply

- Fuel cell – an electrochemical generator

- Power supply – usually a main hook-up

- Photo voltaic device – generates electricity from light

- Thermo electric generator – generates electricity from temperature gradients

- Electrical generator – an electromechanical power source

- Piezoelectric pressure - creates electricity from mechanical strain

- Van de Graaferator - Van de Graaff generator or essentially creating voltage from friction

Passive Components

Components incapable of controlling current by means of another electrical signal are called *passive* devices. Resistors, capacitors, inductors, transformers, and even diodes are all considered passive devices.

Resistors

SMD resistors on a backside of a PCB

Pass current in proportion to voltage (Ohm's law) and oppose current.

- Resistor – fixed value

  o Power resistor – larger to safely dissipate heat generated

  o SIP or DIP resistor network – array of resistors in one package

- Variable resistor

  o Rheostat – two-terminal variable resistor (often for high power)

  o Potentiometer – three-terminal variable resistor (variable voltage divider)

  o Trim pot – Small potentiometer, usually for internal adjustments

  o Thermistor – thermally sensitive resistor whose prime function is to exhibit a large, predictable and precise change in electrical resistance when subjected to a corresponding change in body temperature.

o   Humistor – humidity-varied resistor

o   Photoresistor

o   Memristor

o   Varistor, Voltage Dependent Resistor, MOV – Passes current when excessive voltage is present

- Resistance wire, Nichrome wire – wire of high-resistance material, often used as a heating element

- Heater – heating element

Capacitors

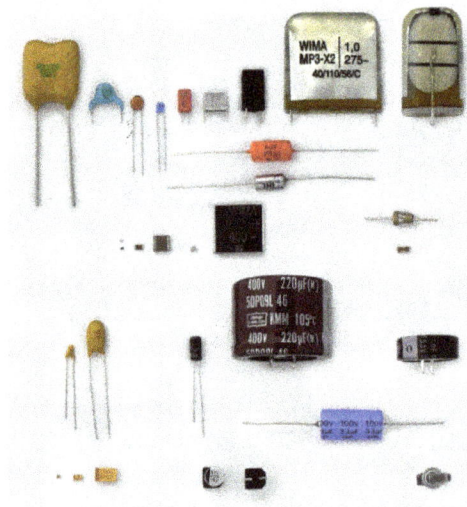

Some different capacitors for electronic equipment

Capacitors store and release electrical charge. They are used for filtering power supply lines, tuning resonant circuits, and for blocking DC voltages while passing AC signals, among numerous other uses.

- Capacitor

  o   Integrated capacitors

      □   MIS capacitor

      □   Trench capacitor

  o   Fixed capacitors

      □   Ceramic capacitor

      □   Film capacitor

      □   Electrolytic capacitor

      ☐ Aluminum electrolytic capacitor

      ☐ Tantalum electrolytic capacitor

      ☐ Niobium electrolytic capacitor

      ☐ Polymer capacitor, OS-CON

    ☐ Supercapacitor (Electric double-layer capacitor)

      ☐ Nanoionic supercapacitor

      ☐ Lithium-ion capacitor

    ☐ Mica capacitor

    ☐ Vacuum capacitor

- o Variable capacitor – adjustable capacitance

    ☐ Tuning capacitor – variable capacitor for tuning a radio, oscillator, or tuned circuit

    ☐ Trim capacitor– small variable capacitor is usually for slight internal adjustments made with a small screw driver turned into the right position.

    ☐ Vacuum variable capacitor

- o Capacitors for special applications

    ☐ Power capacitor

    ☐ Safety capacitor

    ☐ Filter capacitor

    ☐ Light-emitting capacitor

    ☐ Motor capacitor

    ☐ Photoflash capacitor

    ☐ Reservoir capacitor

- o Capacitor network (array)
- • Varicap diode – AC capacitance varies according to the DC voltage applied

Magnetic (Inductive) Devices

Electrical components that use magnetism in the storage and release of electrical charge through current:

- • Inductor, coil, choke

- Variable inductor

- Saturable Inductor

- Transformer

- Magnetic amplifier (toroid)

- ferrite impedances, beads

- Motor / Generator

- Solenoid

- Loudspeaker and microphone

Memristor

Electrical components that pass charge in proportion to magnetism or magnetic flux, and have the ability to retain a previous resistive state, hence the name of Memory plus Resistor.

- Memristor

Networks

Components that use more than one type of passive component:

- RC network – forms an RC circuit, used in snubbers

- LC Network – forms an LC circuit, used in tunable transformers and RFI filters.

Transducers, Sensors, Detectors

1. Transducers generate physical effects when driven by an electrical signal, or vice versa.

2. Sensors (detectors) are transducers that react to environmental conditions by changing their electrical properties or generating an electrical signal.

3. The transducers listed here are single electronic components (as opposed to complete assemblies), and are passive. Only the most common ones are listed here.

- Audio

    o   Loudspeaker – Electromagnetic or piezoelectric device to generate full audio

    o   Buzzer – Electromagnetic or piezoelectric sounder to generate tones

- Position, motion

    o   Linear variable differential transformer (LVDT) – Magnetic – detects linear position

    o   Rotary encoder, Shaft Encoder – Optical, magnetic, resistive or switches – detects absolute or relative angle or rotational speed

- o Inclinometer – Capacitive – detects angle with respect to gravity

- o Motion sensor, Vibration sensor

- o Flow meter – detects flow in liquid or gas

- Force, torque

  - o Strain gauge – Piezoelectric or resistive – detects squeezing, stretching, twisting

  - o Accelerometer – Piezoelectric – detects acceleration, gravity

- Thermal

  - o Thermocouple, thermopile – Wires that generate a voltage proportional to delta temperature

  - o Thermistor – Resistor whose resistance changes with temperature, up PTC or down NTC

  - o Resistance Temperature Detector (RTD) – Wire whose resistance changes with temperature

  - o Bolometer – Device for measuring the power of incident electromagnetic radiation

  - o Thermal cutoff – Switch that is opened or closed when a set temperature is exceeded

- Magnetic field

  - o Magnetometer, Gauss meter

- Humidity

  - o Hygrometer

- Electromagnetic, light

  - o Photo resistor – Light dependent resistor (LDR)

Antennas

Antennas transmit or receive radio waves

- Elemental dipole

- Yagi

- Phased array

- Loop antenna

- Parabolic dish

- Log-periodic dipole array

- Biconical

- Feedhorn

Assemblies, Modules

Multiple electronic components assembled in a device that is in itself used as a component

- Oscillator

- Display devices

  o Liquid crystal display (LCD)

  o Digital voltmeters

- Filter

Prototyping Aids

- Wire-wrap

- Breadboard

Electromechanical

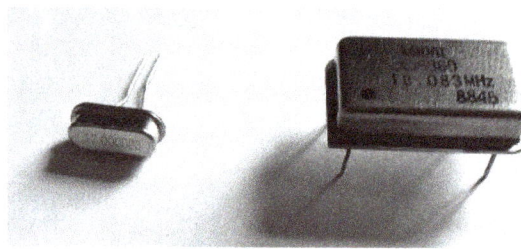

2 crystalline type oscillators

Piezoelectric Devices , Crystals, Resonators

Passive components that use piezoelectric effect:

- Components that use the effect to generate or filter high frequencies

  o Crystal – a ceramic crystal used to generate precise frequencies

  o Ceramic resonator – Is a ceramic crystal used to generate semi-precise frequencies

  o Ceramic filter – Is a ceramic crystal used to filter a band of frequencies such as in radio receivers

  o surface acoustic wave (SAW) filters

- Components that use the effect as mechanical transducers.

     o   Ultrasonic motor – Electric motor that uses the piezoelectric effects

     o   For piezo buzzers and microphones, see the Transducer class below

Terminals and Connectors

Devices to make electrical connection

- Terminal

- Connector

     o   Socket

     o   Screw terminal, Terminal Blocks

     o   Pin header

Cable Assemblies

Cables with connectors or terminals at their ends

- Power cord

- Patch cord

- Test lead

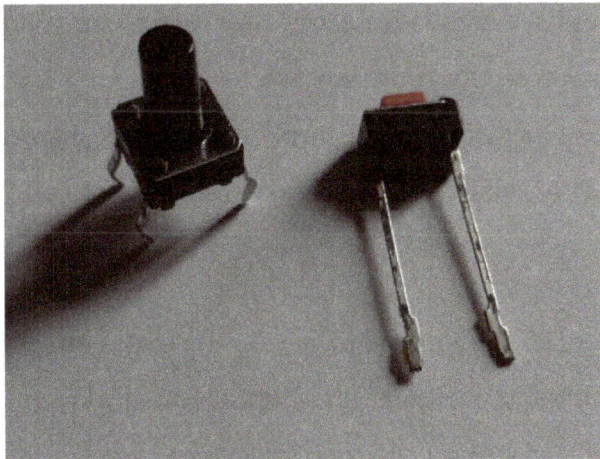

2 different tactile switches

Switches

Components that can pass current ("closed") or break the flow of current ("open"):

- Switch – Manually operated switch.

     o   Electrical description: SPST, SPDT, DPST, DPDT, NPNT (general)

     o   Technology: slide switches, toggle switches, rocker switches, rotary switches, pushbutton switches

- Keypad – Array of pushbutton switches
- DIP switch – Small array of switches for internal configuration settings
- Footswitch – Foot-operated switch
- Knife switch – Switch with unenclosed conductors
- Micro switch – Mechanically activated switch with snap action
- Limit switch – Mechanically activated switch to sense limit of motion
- Mercury switch – Switch sensing tilt
- Centrifugal switch – Switch sensing centrifugal force due to rate of rotation
- Relay – Electrically operated switch
- Reed switch – Magnetically activated switch
- Thermostat – Thermally activated switch
- Humidistat – Humidity activated switch
- Circuit breaker – Switch opened in response to excessive current: a resettable fuse

Protection Devices

Passive components that protect circuits from excessive currents or voltages:

- Fuse – over-current protection, one time use
- Circuit breaker – resettable fuse in the form of a mechanical switch
- Resettable fuse or PolySwitch – circuit breaker action using solid state device
- Ground-fault protection or residual-current device – circuit breaker sensitive to mains currents passing to ground
- Metal oxide varistor (MOV), surge absorber, TVS – Over-voltage protection.
- Inrush current limiter – protection against initial Inrush current
- Gas discharge tube – protection against high voltage surges
- Spark gap – electrodes with a gap to arc over at a high voltage
- Lightning arrester – spark gap used to protect against lightning strikes

Mechanical Accessories

- Enclosure (electrical)
- Heat sink

- Fan

Other

- Printed circuit boards

- Lamp

- Waveguide

- Memristor

Obsolete

- Carbon amplifier

- Carbon arc (negative resistance device)

- Dynamo (historic rf generator)

- Coherer

Standard Symbols

On a circuit diagram, electronic devices are represented by conventional symbols. Reference designators are applied to the symbols to identify the component.

Various Electronics Components

Semiconductor Device

Semiconductor devices are electronic components that exploit the electronic properties of semiconductor materials, principally silicon, germanium, and gallium arsenide, as well as organic semiconductors. Semiconductor devices have replaced thermionic devices (vacuum tubes) in most applications. They use electronic conduction in the solid state as opposed to the gaseous state or thermionic emission in a high vacuum.

Semiconductor devices are manufactured both as single discrete devices and as *integrated circuits* (ICs), which consist of a number—from a few (as low as two) to billions—of devices manufactured and interconnected on a single semiconductor substrate, or wafer.

Semiconductor materials are useful because their behavior can be easily manipulated by the addition of impurities, known as doping. Semiconductor conductivity can be controlled by the introduction of an electric or magnetic field, by exposure to light or heat, or by the mechanical deformation of a doped monocrystalline grid; thus, semiconductors can make excellent sensors. Current conduction in a semiconductor occurs via mobile or "free" *electrons* and *holes*, collectively known as *charge carriers*. Doping a semiconductor such as silicon with a small amount of impurity atoms, such as phosphorus or boron, greatly increases the number of free electrons or holes within the semiconductor. When a doped semiconductor contains excess holes it is called "p-type", and when it contains excess free electrons it is known as "n-type", where p (positive for holes) or n (negative for electrons) is the sign of the charge of the majority mobile charge carriers. The semiconductor material used in devices is doped under highly controlled conditions in a fabrication facility, or *fab*, to control precisely the location and concentration of p- and n-type dopants. The junctions which form where n-type and p-type semiconductors join together are called p–n junctions.

Diode

A semiconductor diode is a device typically made from a single p–n junction. At the junction of a p-type and an n-type semiconductor there forms a depletion region where current conduction is inhibited by the lack of mobile charge carriers. When the device is *forward biased* (connected with the p-side at higher electric potential than the n-side), this depletion region is diminished, allowing for significant conduction, while only very small current can be achieved when the diode is *reverse biased* and thus the depletion region expanded.

Exposing a semiconductor to light can generate electron–hole pairs, which increases the number of free carriers and thereby the conductivity. Diodes optimized to take advantage of this phenomenon are known as *photodiodes*. Compound semiconductor diodes can also be used to generate light, as in light-emitting diodes and laser diodes.

Transistor

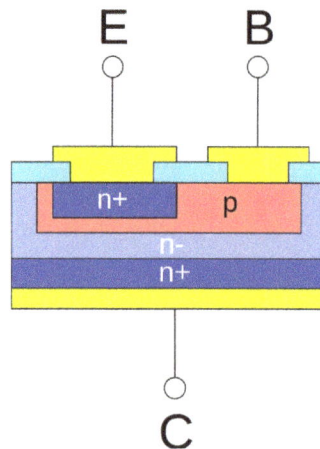

An n–p–n bipolar junction transistor structure

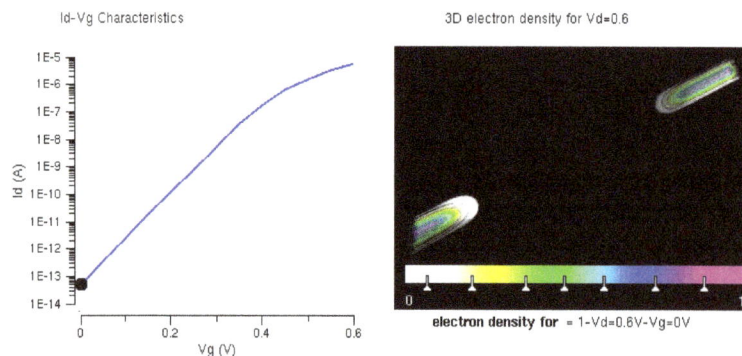

Operation of a MOSFET and its Id-Vg curve. At first, when no gate voltage is applied. There is no inversion electron in the channel, the device is OFF. As gate voltage increase, inversion electron density in the channel increase, current increase, the device turns on.

Bipolar Junction Transistor

Bipolar junction transistors are formed from two p–n junctions, in either n–p–n or p–n–p configuration. The middle, or *base*, region between the junctions is typically very narrow. The other regions, and their associated terminals, are known as the *emitter* and the *collector*. A small current injected through the junction between the base and the emitter changes the properties of the base-collector junction so that it can conduct current even though it is reverse biased. This creates a much larger current between the collector and emitter, controlled by the base-emitter current.

Field-effect Transistor

Another type of transistor, the field-effect transistor, operates on the principle that semiconductor conductivity can be increased or decreased by the presence of an electric field. An electric field can increase the number of free electrons and holes in a semiconductor, thereby changing its conductivity. The field may be applied by a reverse-biased p–n junction, forming a *junction field-effect transistor* (JFET) or by an electrode insulated from the bulk material by an oxide layer, forming a *metal–oxide–semiconductor field-effect transistor* (MOSFET).

The MOSFET, a solid-state device, is the most used semiconductor device today. The *gate* electrode is charged to produce an electric field that controls the conductivity of a "channel" between two terminals, called the *source* and *drain*. Depending on the type of carrier in the channel, the device may be an *n-channel* (for electrons) or a *p-channel* (for holes) MOSFET. Although the MOSFET is named in part for its "metal" gate, in modern devices polysilicon is typically used instead.

Semiconductor Device Materials

By far, silicon (Si) is the most widely used material in semiconductor devices. Its combination of low raw material cost, relatively simple processing, and a useful temperature range makes it currently the best compromise among the various competing materials. Silicon used in semiconductor device manufacturing is currently fabricated into boules that are large enough in diameter to allow the production of 300 mm (12 in.) wafers.

Germanium (Ge) was a widely used early semiconductor material but its thermal sensitivity makes it less useful than silicon. Today, germanium is often alloyed with silicon for use in very-high-speed SiGe devices; IBM is a major producer of such devices.

Gallium arsenide (GaAs) is also widely used in high-speed devices but so far, it has been difficult to form large-diameter boules of this material, limiting the wafer diameter to sizes significantly smaller than silicon wafers thus making mass production of GaAs devices significantly more expensive than silicon.

Other less common materials are also in use or under investigation.

Silicon carbide (SiC) has found some application as the raw material for blue light-emitting diodes (LEDs) and is being investigated for use in semiconductor devices that could withstand very high operating temperatures and environments with the presence of significant levels of ionizing radiation. IMPATT diodes have also been fabricated from SiC.

Various indium compounds (indium arsenide, indium antimonide, and indium phosphide) are also being used in LEDs and solid state laser diodes. Selenium sulfide is being studied in the manufacture of photovoltaic solar cells.

The most common use for organic semiconductors is Organic light-emitting diodes.

List of Common Semiconductor Devices

Two-terminal devices:

- DIAC
- Diode (rectifier diode)
- Gunn diode
- IMPATT diode
- Laser diode
- Light-emitting diode (LED)
- Photocell
- Phototransistor
- PIN diode
- Schottky diode
- Solar cell
- Transient-voltage-suppression diode
- Tunnel diode
- VCSEL
- Zener diode

Three-terminal devices:

- Bipolar transistor
- Darlington transistor
- Field-effect transistor
- Insulated-gate bipolar transistor (IGBT)
- Silicon-controlled rectifier

- Thyristor

- TRIAC

- Unijunction transistor

Four-terminal devices:

- Hall effect sensor (magnetic field sensor)

- Photocoupler (Optocoupler)

Semiconductor Device Applications

All transistor types can be used as the building blocks of logic gates, which are fundamental in the design of digital circuits. In digital circuits like microprocessors, transistors act as on-off switches; in the MOSFET, for instance, the voltage applied to the gate determines whether the switch is on or off.

Transistors used for analog circuits do not act as on-off switches; rather, they respond to a continuous range of inputs with a continuous range of outputs. Common analog circuits include amplifiers and oscillators.

Circuits that interface or translate between digital circuits and analog circuits are known as mixed-signal circuits.

Power semiconductor devices are discrete devices or integrated circuits intended for high current or high voltage applications. Power integrated circuits combine IC technology with power semiconductor technology, these are sometimes referred to as "smart" power devices. Several companies specialize in manufacturing power semiconductors.

Component Identifiers

The type designators of semiconductor devices are often manufacturer specific. Nevertheless, there have been attempts at creating standards for type codes, and a subset of devices follow those. For discrete devices, for example, there are three standards: JEDEC JESD370B in United States, Pro Electron in Europe and Japanese Industrial Standards (JIS) in Japan.

History of Semiconductor Device Development

Cat's-Whisker Detector

Semiconductors had been used in the electronics field for some time before the invention of the transistor. Around the turn of the 20th century they were quite common as detectors in radios, used in a device called a "cat's whisker" developed by Jagadish Chandra Bose and others. These detectors were somewhat troublesome, however, requiring the operator to move a small tungsten filament (the whisker) around the surface of a galena (lead sulfide) or carborundum (silicon carbide) crystal until it suddenly started working. Then, over a period of a few hours or days, the cat's whisker would slowly stop working and the process would have to be repeated. At the time their operation was completely mysterious. After the introduction of the more reliable and amplified vacuum tube based radios, the cat's whisker systems quickly disappeared. The "cat's whisker" is a primitive example of a special type of diode still popular today, called a Schottky diode.

Metal Rectifier

Another early type of semiconductor device is the metal rectifier in which the semiconductor is copper oxide or selenium. Westinghouse Electric (1886) was a major manufacturer of these rectifiers.

World War II

During World War II, radar research quickly pushed radar receivers to operate at ever higher frequencies and the traditional tube based radio receivers no longer worked well. The introduction of the cavity magnetron from Britain to the United States in 1940 during the Tizard Mission resulted in a pressing need for a practical high-frequency amplifier.

On a whim, Russell Ohl of Bell Laboratories decided to try a cat's whisker. By this point they had not been in use for a number of years, and no one at the labs had one. After hunting one down at a used radio store in Manhattan, he found that it worked much better than tube-based systems.

Ohl investigated why the cat's whisker functioned so well. He spent most of 1939 trying to grow more pure versions of the crystals. He soon found that with higher quality crystals their finicky behaviour went away, but so did their ability to operate as a radio detector. One day he found one of his purest crystals nevertheless worked well, and interestingly, it had a clearly visible crack near the middle. However as he moved about the room trying to test it, the detector would mysteriously work, and then stop again. After some study he found that the behaviour was controlled by the light in the room—more light caused more conductance in the crystal. He invited several other people to see this crystal, and Walter Brattain immediately realized there was some sort of junction at the crack.

Further research cleared up the remaining mystery. The crystal had cracked because either side contained very slightly different amounts of the impurities Ohl could not remove—about 0.2%. One side of the crystal had impurities that added extra electrons (the carriers of electric current) and made it a "conductor". The other had impurities that wanted to bind to these electrons, making it (what he called) an "insulator". Because the two parts of the crystal were in contact with each other, the electrons could be pushed out of the conductive side which had extra electrons (soon to be known as the *emitter*) and replaced by new ones being provided (from a battery, for instance) where they would flow into the insulating portion and be collected by the whisker filament (named the *collector*). However, when the voltage was reversed the electrons being pushed into the collector would quickly fill up the "holes" (the electron-needy impurities), and conduction would stop almost instantly. This junction of the two crystals (or parts of one crystal) created a solid-state diode, and the concept soon became known as semiconduction. The mechanism of action when the diode is off has to do with the separation of charge carriers around the junction. This is called a "depletion region".

Development of the Diode

Armed with the knowledge of how these new diodes worked, a vigorous effort began to learn how to build them on demand. Teams at Purdue University, Bell Labs, MIT, and the University of Chicago all joined forces to build better crystals. Within a year germanium production had been perfected to the point where military-grade diodes were being used in most radar sets.

Development of the Transistor

After the war, William Shockley decided to attempt the building of a triode-like semiconductor device. He secured funding and lab space, and went to work on the problem with Brattain and John Bardeen.

The key to the development of the transistor was the further understanding of the process of the electron mobility in a semiconductor. It was realized that if there were some way to control the flow of the electrons from the emitter to the collector of this newly discovered diode, an amplifier could be built. For instance, if contacts are placed on both sides of a single type of crystal, current will not flow between them through the crystal. However if a third contact could then "inject" electrons or holes into the material, current would flow.

Actually doing this appeared to be very difficult. If the crystal were of any reasonable size, the number of electrons (or holes) required to be injected would have to be very large, making it less than useful as an amplifier because it would require a large injection current to start with. That said, the whole idea of the crystal diode was that the crystal itself could provide the electrons over a very small distance, the depletion region. The key appeared to be to place the input and output contacts very close together on the surface of the crystal on either side of this region.

Brattain started working on building such a device, and tantalizing hints of amplification continued to appear as the team worked on the problem. Sometimes the system would work but then stop working unexpectedly. In one instance a non-working system started working when placed in water. Ohl and Brattain eventually developed a new branch of quantum mechanics, which became known as surface physics, to account for the behaviour. The electrons in any one piece of the crystal would migrate about due to nearby charges. Electrons in the emitters, or the "holes" in the collectors, would cluster at the surface of the crystal where they could find their opposite charge "floating around" in the air (or water). Yet they could be pushed away from the surface with the application of a small amount of charge from any other location on the crystal. Instead of needing a large supply of injected electrons, a very small number in the right place on the crystal would accomplish the same thing.

Their understanding solved the problem of needing a very small control area to some degree. Instead of needing two separate semiconductors connected by a common, but tiny, region, a single larger surface would serve. The electron-emitting and collecting leads would both be placed very close together on the top, with the control lead placed on the base of the crystal. When current flowed through this "base" lead, the electrons or holes would be pushed out, across the block of semiconductor, and collect on the far surface. As long as the emitter and collector were very close together, this should allow enough electrons or holes between them to allow conduction to start.

The First Transistor

A stylized replica of the first transistor

The Bell team made many attempts to build such a system with various tools, but generally failed. Setups where the contacts were close enough were invariably as fragile as the original cat's whisker detectors had been, and would work briefly, if at all. Eventually they had a practical breakthrough. A piece of gold foil was glued to the edge of a plastic wedge, and then the foil was sliced with a razor at the tip of the triangle. The result was two very closely spaced contacts of gold. When the wedge was pushed down onto the surface of a crystal and voltage applied to the other side (on the base of the crystal), current started to flow from one contact to the other as the base voltage pushed the electrons away from the base towards the other side near the contacts. The point-contact transistor had been invented.

While the device was constructed a week earlier, Brattain's notes describe the first demonstration to higher-ups at Bell Labs on the afternoon of 23 December 1947, often given as the birthdate of the transistor. what is now known as the "p–n–p point-contact germanium transistor" operated as a speech amplifier with a power gain of 18 in that trial. John Bardeen, Walter Houser Brattain, and William Bradford Shockley were awarded the 1956 Nobel Prize in physics for their work.

Origin of the Term "Transistor"

Bell Telephone Laboratories needed a generic name for their new invention: "Semiconductor Triode", "Solid Triode", "Surface States Triode" [sic], "Crystal Triode" and "Iotatron" were all considered, but "transistor", coined by John R. Pierce, won an internal ballot. The rationale for the name is described in the following extract from the company's Technical Memoranda (May 28, 1948) calling for votes:

Transistor. This is an abbreviated combination of the words "transconductance" or "transfer", and "varistor". The device logically belongs in the varistor family, and has the transconductance or transfer impedance of a device having gain, so that this combination is descriptive.

Improvements in Transistor Design

Shockley was upset about the device being credited to Brattain and Bardeen, who he felt had built it "behind his back" to take the glory. Matters became worse when Bell Labs lawyers found that some of Shockley's own writings on the transistor were close enough to those of an earlier 1925 patent by Julius Edgar Lilienfeld that they thought it best that his name be left off the patent application.

Shockley was incensed, and decided to demonstrate who was the real brains of the operation. A few months later he invented an entirely new, considerably more robust, type of transistor with a layer or 'sandwich' structure. This structure went on to be used for the vast majority of all transistors into the 1960s, and evolved into the bipolar junction transistor.

With the fragility problems solved, a remaining problem was purity. Making germanium of the required purity was proving to be a serious problem, and limited the yield of transistors that actually worked from a given batch of material. Germanium's sensitivity to temperature also limited its usefulness. Scientists theorized that silicon would be easier to fabricate, but few investigated this possibility. Gordon K. Teal was the first to develop a working silicon transistor, and his company, the nascent Texas Instruments, profited from its technological edge. From the late 1960s most transistors were silicon-based. Within a few years transistor-based products, most notably easily portable radios, were appearing on the market.

A major improvement in manufacturing yield came when a chemist advised the companies fabricating semiconductors to use distilled rather than tap water: calcium ions present in tap water were the cause of the poor yields. "Zone melting", a technique using a band of molten material moving through the crystal, further increased crystal purity.

Resistor

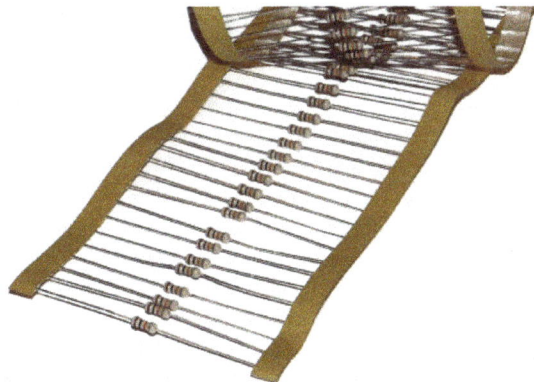

Axial-lead resistors on tape. The component is cut from the
tape during assembly and the part is inserted into the board.

A resistor is a passive two-terminal electrical component that implements electrical resistance as a circuit element. In electronic circuits, resistors are used to reduce current flow, adjust signal levels, to divide voltages, bias active elements, and terminate transmission lines, among other uses. High-power resistors that can dissipate many watts of electrical power as heat may be used as part of motor controls, in power distribution systems, or as test loads for generators. Fixed resistors have resistances that only change slightly with temperature, time or operating voltage. Variable resistors can be used to adjust circuit elements (such as a volume control or a lamp dimmer), or as sensing devices for heat, light, humidity, force, or chemical activity.

Resistors are common elements of electrical networks and electronic circuits and are ubiquitous in electronic equipment. Practical resistors as discrete components can be composed of various compounds and forms. Resistors are also implemented within integrated circuits.

The electrical function of a resistor is specified by its resistance: common commercial resistors are manufactured over a range of more than nine orders of magnitude. The nominal value of the resistance falls within the manufacturing tolerance, indicated on the component.

Electronic Symbols and Notation

Two typical schematic diagram symbols are as follows:

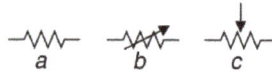

(a) resistor, (b) rheostat (variable resistor), and (c) potentiometer

IEC resistor symbol

The notation to state a resistor's value in a circuit diagram varies. The European notation BS 1852 avoids using a decimal separator, and replaces the decimal separator with the SI prefix symbol for the particular value. For example, 8k2 in a circuit diagram indicates a resistor value of 8.2 kΩ. Additional zeros imply tighter tolerance, for example 15M0. When the value can be expressed without the need for an SI prefix, an "R" is used instead of the decimal separator. For example, 1R2 indicates 1.2 Ω, and 18R indicates 18 Ω. The use of a SI prefix symbol or the letter "R" circumvents the problem that decimal separators tend to "disappear" when photocopying a printed circuit diagram.

Theory of Operation

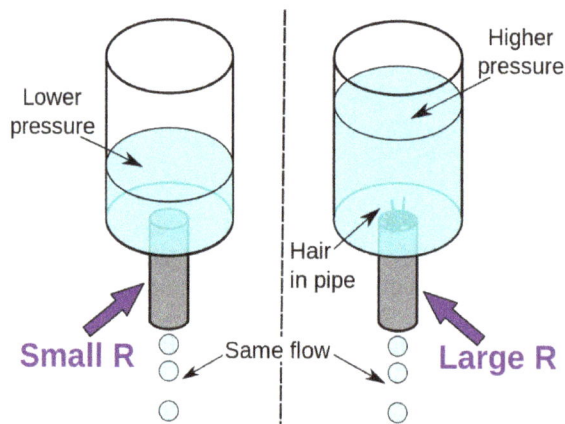

The hydraulic analogy compares electric current flowing through circuits to water flowing through pipes. When a pipe (left) is clogged with hair (right), it takes a larger pressure to achieve the same flow of water. Pushing electric current through a large resistance is like pushing water through a pipe clogged with hair: It requires a larger push (voltage) to drive the same flow (electric current).

Ohm's Law

The behaviour of an ideal resistor is dictated by the relationship specified by Ohm's law:

$$V = I \cdot R.$$

Ohm's law states that the voltage (V) across a resistor is proportional to the current (I), where the constant of proportionality is the resistance (R). For example, if a 300 ohm resistor is attached across the terminals of a 12 volt battery, then a current of 12 / 300 = 0.04 amperes flows through that resistor.

Practical resistors also have some inductance and capacitance which affect the relation between voltage and current in alternating current circuits.

The ohm (symbol: Ω) is the SI unit of electrical resistance, named after Georg Simon Ohm. An ohm is equivalent to a volt per ampere. Since resistors are specified and manufactured over a very large range of values, the derived units of milliohm (1 mΩ = 10^{-3} Ω), kilohm (1 kΩ = 10^{3} Ω), and megohm (1 MΩ = 10^{6} Ω) are also in common usage.

Series and Parallel Resistors

The total resistance of resistors connected in series is the sum of their individual resistance values.

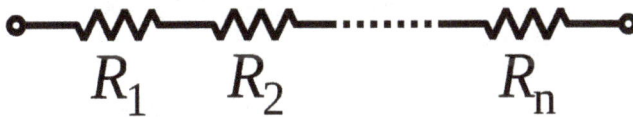

$$R_{eq} = R_1 + R_2 + \cdots + R_n.$$

The total resistance of resistors connected in parallel is the reciprocal of the sum of the reciprocals of the individual resistors.

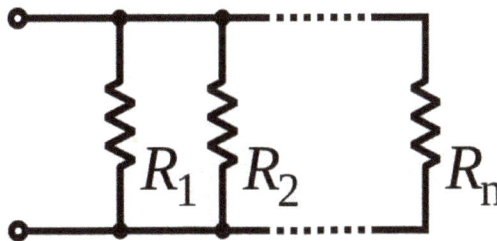

$$\frac{1}{R_{eq}} = \frac{1}{R_1} + \frac{1}{R_2} + \cdots + \frac{1}{R_n}.$$

For example, a 10 ohm resistor connected in parallel with a 5 ohm resistor and a 15 ohm resistor produces 1/1/10 + 1/5 + 1/15 ohms of resistance, or 30/11 = 2.727 ohms.

A resistor network that is a combination of parallel and series connections can be broken up into smaller parts that are either one or the other. Some complex networks of resistors cannot be resolved in this manner, requiring more sophisticated circuit analysis. Generally, the Y-Δ transform, or matrix methods can be used to solve such problems.

Power Dissipation

At any instant, the power P (watts) consumed by a resistor of resistance R (ohms) is calculated as: $P = I^2 R = IV = \dfrac{V^2}{R}$ where V (volts) is the voltage across the resistor and I (amps) is the current flowing through it. Using Ohm's law, the two other forms can be derived. This power is converted into heat which must be dissipated by the resistor's package before its temperature rises excessively.

Resistors are rated according to their maximum power dissipation. Discrete resistors in solid-state electronic systems are typically rated as 1/10, 1/8, or 1/4 watt. They usually absorb much less than a watt of electrical power and require little attention to their power rating.

An aluminium-housed power resistor rated for 50 W when heat-sinked

Resistors required to dissipate substantial amounts of power, particularly used in power supplies, power conversion circuits, and power amplifiers, are generally referred to as *power resistors*; this designation is loosely applied to resistors with power ratings of 1 watt or greater. Power resistors are physically larger and may not use the preferred values, color codes, and external packages described below.

VZR power resistor 1.5kΩ 12W, manufactured in 1963 in the Soviet Union

If the average power dissipated by a resistor is more than its power rating, damage to the resistor may occur, permanently altering its resistance; this is distinct from the reversible change in resistance due to its temperature coefficient when it warms. Excessive power dissipation may raise the temperature of the resistor to a point where it can burn the circuit board or adjacent components, or even cause a fire. There are flameproof resistors that fail (open circuit) before they overheat dangerously.

Since poor air circulation, high altitude, or high operating temperatures may occur, resistors may be specified with higher rated dissipation than is experienced in service.

All resistors have a maximum voltage rating; this may limit the power dissipation for higher resistance values.

Nonideal Properties

Practical resistors have a series inductance and a small parallel capacitance; these specifications can be important in high-frequency applications. In a low-noise amplifier or pre-amp, the noise characteristics of a resistor may be an issue.

The temperature coefficient of the resistance may also be of concern in some precision applications.

The unwanted inductance, excess noise, and temperature coefficient are mainly dependent on the technology used in manufacturing the resistor. They are not normally specified individually for a particular family of resistors manufactured using a particular technology. A family of discrete resistors is also characterized according to its form factor, that is, the size of the device and the position of its leads (or terminals) which is relevant in the practical manufacturing of circuits using them.

Practical resistors are also specified as having a maximum power rating which must exceed the anticipated power dissipation of that resistor in a particular circuit: this is mainly of concern in power electronics applications. Resistors with higher power ratings are physically larger and may require heat sinks. In a high-voltage circuit, attention must sometimes be paid to the rated maximum working voltage of the resistor. While there is no minimum working voltage for a given resistor, failure to account for a resistor's maximum rating may cause the resistor to incinerate when current is run through it.

Fixed Resistor

A single in line (SIL) resistor package with 8 individual, 47 ohm resistors.
One end of each resistor is connected to a separate pin and the other ends are all connected
together to the remaining (common) pin – pin 1, at the end identified by the white dot.

Lead Arrangements

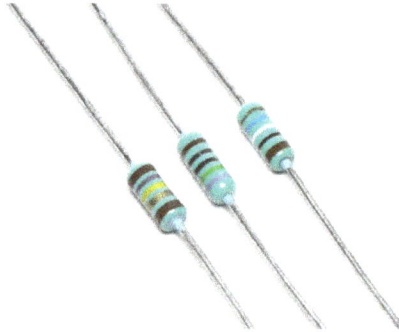

Resistors with wire leads for through-hole mounting

Through-hole components typically have "leads" (pronounced \ˈlēdz\) leaving the body "axially," that is, on a line parallel with the part's longest axis. Others have leads coming off their body "radially" instead. Other components may be SMT (surface mount technology), while high power resistors may have one of their leads designed into the heat sink.

Carbon Composition

Three carbon composition resistors in a 1960s valve (vacuum tube) radio

Carbon composition resistors consist of a solid cylindrical resistive element with embedded wire leads or metal end caps to which the lead wires are attached. The body of the resistor is protected with paint or plastic. Early 20th-century carbon composition resistors had uninsulated bodies; the lead wires were wrapped around the ends of the resistance element rod and soldered. The completed resistor was painted for color-coding of its value.

The resistive element is made from a mixture of finely powdered carbon and an insulating material, usually ceramic. A resin holds the mixture together. The resistance is determined by the ratio of the fill material (the powdered ceramic) to the carbon. Higher concentrations of carbon, which is a good conductor, result in lower resistance. Carbon composition resistors were commonly used in the 1960s and earlier, but are not popular for general use now as other types have better specifications, such as tolerance, voltage dependence, and stress. Carbon composition resistors change value when stressed with over-voltages. Moreover, if internal moisture content, from exposure for some length of time to a humid environment, is significant, soldering heat creates a non-reversible change in resistance value. Carbon composition resistors have poor stability with time and were consequently factory sorted to, at best, only 5% tolerance. These resistors, however, if never subjected to overvoltage nor overheating were remarkably reliable considering the component's size.

Carbon composition resistors are still available, but comparatively quite costly. Values ranged from fractions of an ohm to 22 megohms. Due to their high price, these resistors are no longer used in most applications. However, they are used in power supplies and welding controls.

Carbon Pile

A carbon pile resistor is made of a stack of carbon disks compressed between two metal contact plates. Adjusting the clamping pressure changes the resistance between the plates. These resistors are used when an adjustable load is required, for example in testing automotive batteries or radio transmitters. A carbon pile resistor can also be used as a speed control for small motors in household appliances (sewing machines, hand-held mixers) with ratings up to a few hundred watts. A carbon pile resistor can be incorporated in automatic voltage regulators for generators, where the carbon pile controls the field current to maintain relatively constant voltage. The principle is also applied in the carbon microphone.

Carbon Film

Carbon film resistor with exposed carbon spiral (Tesla TR-212 1 kΩ)

A carbon film is deposited on an insulating substrate, and a helix is cut in it to create a long, narrow resistive path. Varying shapes, coupled with the resistivity of amorphous carbon (ranging from 500 to 800 µΩ m), can provide a wide range of resistance values. Compared to carbon composition they feature low noise, because of the precise distribution of the pure graphite without binding. Carbon film resistors feature a power rating range of 0.125 W to 5 W at 70 °C. Resistances available range from 1 ohm to 10 megohm. The carbon film resistor has an operating temperature range of −55 °C to 155 °C. It has 200 to 600 volts maximum working voltage range. Special carbon film resistors are used in applications requiring high pulse stability.

Printed Carbon Resistor

Carbon composition resistors can be printed directly onto printed circuit board (PCB) substrates as part of the PCB manufacturing process. Although this technique is more common on hybrid PCB modules, it can also be used on standard fibreglass PCBs. Tolerances are typically quite large, and can be in the order of 30%. A typical application would be non-critical pull-up resistors.

A carbon resistor printed directly onto the SMD pads on a PCB. Inside a 1989 vintage Psion II Organiser

Thick and Thin Film

Laser Trimmed Precision Thin Film Resistor Network from Fluke, used in the
Keithley DMM7510 multimeter. Ceramic backed with glass hermetic seal cover.

Thick film resistors became popular during the 1970s, and most SMD (surface mount device) resistors today are of this type. The resistive element of thick films is 1000 times thicker than thin films, but the principal difference is how the film is applied to the cylinder (axial resistors) or the surface (SMD resistors).

Thin film resistors are made by sputtering (a method of vacuum deposition) the resistive material onto an insulating substrate. The film is then etched in a similar manner to the old (subtractive) process for making printed circuit boards; that is, the surface is coated with a photo-sensitive material, then covered by a pattern film, irradiated with ultraviolet light, and then the exposed photo-sensitive coating is developed, and underlying thin film is etched away.

Thick film resistors are manufactured using screen and stencil printing processes.

Because the time during which the sputtering is performed can be controlled, the thickness of the thin film can be accurately controlled. The type of material is also usually different consisting of one or more ceramic (cermet) conductors such as tantalum nitride (TaN), ruthenium oxide (RuO_2), lead oxide (PbO), bismuth ruthenate ($Bi_2Ru_2O_7$), nickel chromium (NiCr), or bismuth iridate ($Bi_2Ir_2O_7$).

The resistance of both thin and thick film resistors after manufacture is not highly accurate; they are usually trimmed to an accurate value by abrasive or laser trimming. Thin film resistors are usually specified with tolerances of 0.1, 0.2, 0.5, or 1%, and with temperature coefficients of 5 to 25 ppm/K. They also have much lower noise levels, on the level of 10–100 times less than thick film resistors.

Thick film resistors may use the same conductive ceramics, but they are mixed with sintered (powdered) glass and a carrier liquid so that the composite can be screen-printed. This composite of glass and conductive ceramic (cermet) material is then fused (baked) in an oven at about 850 °C.

Thick film resistors, when first manufactured, had tolerances of 5%, but standard tolerances have improved to 2% or 1% in the last few decades. Temperature coefficients of thick film resistors are high, typically ±200 or ±250 ppm/K; a 40 kelvin (70 °F) temperature change can change the resistance by 1%.

Thin film resistors are usually far more expensive than thick film resistors. For example, SMD thin film resistors, with 0.5% tolerances, and with 25 ppm/K temperature coefficients, when bought in full size reel quantities, are about twice the cost of 1%, 250 ppm/K thick film resistors.

Metal Film

A common type of axial-leaded resistor today is the metal-film resistor. Metal Electrode Leadless Face (MELF) resistors often use the same technology, and are also cylindrically shaped but are designed for surface mounting. Note that other types of resistors (e.g., carbon composition) are also available in MELF packages.

Metal film resistors are usually coated with nickel chromium (NiCr), but might be coated with any of the cermet materials listed above for thin film resistors. Unlike thin film resistors, the material may be applied using different techniques than sputtering (though this is one of the techniques). Also, unlike thin-film resistors, the resistance value is determined by cutting a helix through the coating rather than by etching. (This is similar to the way carbon resistors are made.) The result is a reasonable tolerance (0.5%, 1%, or 2%) and a temperature coefficient that is generally between 50 and 100 ppm/K. Metal film resistors possess good noise characteristics and low non-linearity due to a low voltage coefficient. Also beneficial are their tight tolerance, low temperature coefficient and long-term stability.

Metal Oxide Film

Metal-oxide film resistors are made of metal oxides which results in a higher operating temperature and greater stability/reliability than Metal film. They are used in applications with high endurance demands.

Wire Wound

High-power wire wound resistors used for dynamic braking on an electric railway car.
Such resistors may dissipate many kilowatts for an extended length of time.

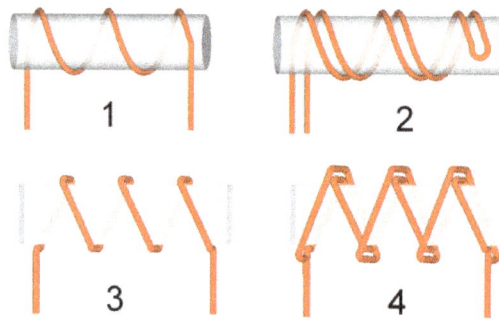

Types of windings in wire resistors:
1   common   2   bifilar   3   common on a thin former   4   Ayrton-Perry

Wirewound resistors are commonly made by winding a metal wire, usually nichrome, around a ceramic, plastic, or fiberglass core. The ends of the wire are soldered or welded to two caps or rings, attached to the ends of the core. The assembly is protected with a layer of paint, molded plastic, or an enamel coating baked at high temperature. These resistors are designed to withstand unusually high temperatures of up to 450 °C. Wire leads in low power wirewound resistors are usually between 0.6 and 0.8 mm in diameter and tinned for ease of soldering. For higher power wirewound resistors, either a ceramic outer case or an aluminum outer case on top of an insulating layer is used – if the outer case is ceramic, such resistors are sometimes described as "cement" resistors, though they do not actually contain any traditional cement. The aluminum-cased types are designed to be attached to a heat sink to dissipate the heat; the rated power is dependent on being used with a suitable heat sink, e.g., a 50 W power rated resistor overheats at a fraction of the power dissipation if not used with a heat sink. Large wirewound resistors may be rated for 1,000 watts or more.

Because wirewound resistors are coils they have more undesirable inductance than other types of resistor, although winding the wire in sections with alternately reversed direction can minimize inductance. Other techniques employ bifilar winding, or a flat thin former (to reduce cross-section area of the coil). For the most demanding circuits, resistors with Ayrton-Perry winding are used.

Applications of wirewound resistors are similar to those of composition resistors with the exception of the high frequency. The high frequency response of wirewound resistors is substantially worse than that of a composition resistor.

Foil Resistor

The primary resistance element of a foil resistor is a special alloy foil several micrometers thick. Since their introduction in the 1960s, foil resistors have had the best precision and stability of any resistor available. One of the important parameters influencing stability is the temperature coefficient of resistance (TCR). The TCR of foil resistors is extremely low, and has been further improved over the years. One range of ultra-precision foil resistors offers a TCR of 0.14 ppm/°C, tolerance ±0.005%, long-term stability (1 year) 25 ppm, (3 years) 50 ppm (further improved 5-fold by hermetic sealing), stability under load (2000 hours) 0.03%, thermal EMF 0.1 μV/°C, noise −42 dB, voltage coefficient 0.1 ppm/V, inductance 0.08 μH, capacitance 0.5 pF.

Ammeter Shunts

An ammeter shunt is a special type of current-sensing resistor, having four terminals and a value in milliohms or even micro-ohms. Current-measuring instruments, by themselves, can usually accept only limited currents. To measure high currents, the current passes through the shunt across which the voltage drop is measured and interpreted as current. A typical shunt consists of two solid metal blocks, sometimes brass, mounted on an insulating base. Between the blocks, and soldered or brazed to them, are one or more strips of low temperature coefficient of resistance (TCR) manganin alloy. Large bolts threaded into the blocks make the current connections, while much smaller screws provide volt meter connections. Shunts are rated by full-scale current, and often have a voltage drop of 50 mV at rated current. Such meters are adapted to the shunt full current rating by using an appropriately marked dial face; no change need to be made to the other parts of the meter.

Grid Resistor

In heavy-duty industrial high-current applications, a grid resistor is a large convection-cooled lattice of stamped metal alloy strips connected in rows between two electrodes. Such industrial grade resistors can be as large as a refrigerator; some designs can handle over 500 amperes of current, with a range of resistances extending lower than 0.04 ohms. They are used in applications such as dynamic braking and load banking for locomotives and trams, neutral grounding for industrial AC distribution, control loads for cranes and heavy equipment, load testing of generators and harmonic filtering for electric substations.

The term *grid resistor* is sometimes used to describe a resistor of any type connected to the control grid of a vacuum tube. This is not a resistor technology; it is an electronic circuit topology.

Special Varieties

- Cermet
- Phenolic
- Tantalum
- Water resistor

Variable Resistors

Adjustable Resistors

A resistor may have one or more fixed tapping points so that the resistance can be changed by moving the connecting wires to different terminals. Some wirewound power resistors have a tapping point that can slide along the resistance element, allowing a larger or smaller part of the resistance to be used.

Where continuous adjustment of the resistance value during operation of equipment is required, the sliding resistance tap can be connected to a knob accessible to an operator. Such a device is called a rheostat and has two terminals.

Potentiometers

Typical panel mount potentiometer

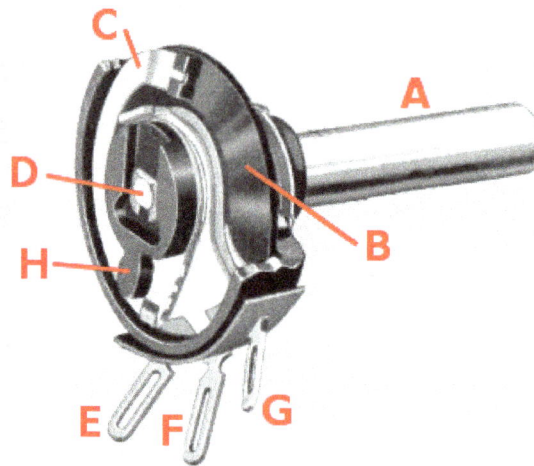

Drawing of potentiometer with case cut away, showing parts: (A) shaft, (B) stationary carbon composition resistance element, (C) phosphor bronze wiper, (D) shaft attached to wiper, (E, G) terminals connected to ends of resistance element, (F) terminal connected to wiper.

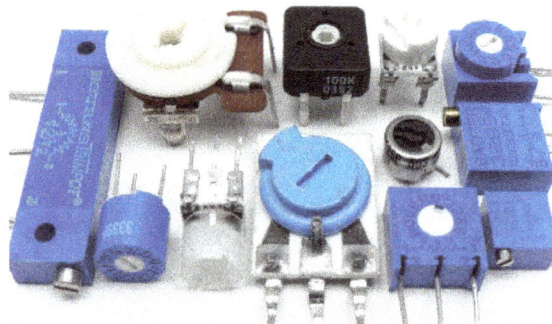

An assortment of small through-hole potentiometers designed for mounting on printed circuit boards.

A potentiometer or *pot* is a three-terminal resistor with a continuously adjustable tapping point controlled by rotation of a shaft or knob or by a linear slider. It is called a potentiometer because it can be connected as an adjustable voltage divider to provide a variable potential at the terminal connected to the tapping point. A volume control for an audio device is a common use of a potentiometer. A typical low power potentiometer is constructed of a flat resistance element *(B)* of carbon composition, metal film, or conductive plastic, with a springy phosphor bronze wiper contact *(C)* which moves along the surface. An alternate construction is resistance wire wound on a form, with the wiper sliding axially along the coil. These have lower resolution, since as the wiper moves the resistance changes in steps equal to the resistance of a single turn.

High-resolution multiturn potentiometers are used in a few precision applications. These have wirewound resistance elements typically wound on a helical mandrel, with the wiper moving on a helical track as the control is turned, making continuous contact with the wire. Some include a conductive-plastic resistance coating over the wire to improve resolution. These typically offer ten turns of their shafts to cover their full range. They are usually set with dials that include a simple turns counter and a graduated dial, and can typically achieve three digit resolution. Electronic analog computers used them in quantity for setting coefficients, and delayed-sweep oscilloscopes of recent decades included one on their panels.

Resistance Decade Boxes

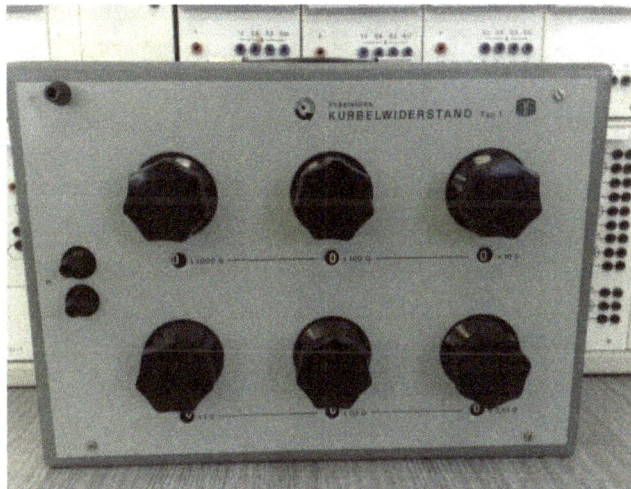

Resistance decade box "Kurbelwiderstand", made in former East Germany.

A resistance decade box or resistor substitution box is a unit containing resistors of many values, with one or more mechanical switches which allow any one of various discrete resistances offered by the box to be dialed in. Usually the resistance is accurate to high precision, ranging from laboratory/calibration grade accuracy of 20 parts per million, to field grade at 1%. Inexpensive boxes with lesser accuracy are also available. All types offer a convenient way of selecting and quickly changing a resistance in laboratory, experimental and development work without needing to attach resistors one by one, or even stock each value. The range of resistance provided, the maximum resolution, and the accuracy characterize the box. For example, one box offers resistances from 0 to 100 megohms, maximum resolution 0.1 ohm, accuracy 0.1%.

Special Devices

There are various devices whose resistance changes with various quantities. The resistance of NTC thermistors exhibit a strong negative temperature coefficient, making them useful for measuring temperatures. Since their resistance can be large until they are allowed to heat up due to the passage of current, they are also commonly used to prevent excessive current surges when equipment is powered on. Similarly, the resistance of a humistor varies with humidity. One sort of photodetector, the photoresistor, has a resistance which varies with illumination.

The strain gauge, invented by Edward E. Simmons and Arthur C. Ruge in 1938, is a type of resistor that changes value with applied strain. A single resistor may be used, or a pair (half bridge), or four resistors connected in a Wheatstone bridge configuration. The strain resistor is bonded with adhesive to an object that is subjected to mechanical strain. With the strain gauge and a filter, amplifier, and analog/digital converter, the strain on an object can be measured.

A related but more recent invention uses a Quantum Tunnelling Composite to sense mechanical stress. It passes a current whose magnitude can vary by a factor of 10^{12} in response to changes in applied pressure.

Measurement

The value of a resistor can be measured with an ohmmeter, which may be one function of a multimeter. Usually, probes on the ends of test leads connect to the resistor. A simple ohmmeter may apply a voltage from a battery across the unknown resistor (with an internal resistor of a known value in series) producing a current which drives a meter movement. The current, in accordance with Ohm's law, is inversely proportional to the sum of the internal resistance and the resistor being tested, resulting in an analog meter scale which is very non-linear, calibrated from infinity to 0 ohms. A digital multimeter, using active electronics, may instead pass a specified current through the test resistance. The voltage generated across the test resistance in that case is linearly proportional to its resistance, which is measured and displayed. In either case the low-resistance ranges of the meter pass much more current through the test leads than do high-resistance ranges, in order for the voltages present to be at reasonable levels (generally below 10 volts) but still measurable.

Measuring low-value resistors, such as fractional-ohm resistors, with acceptable accuracy requires four-terminal connections. One pair of terminals applies a known, calibrated current to the resistor, while the other pair senses the voltage drop across the resistor. Some laboratory quality ohmmeters, especially milliohmmeters, and even some of the better digital multimeters sense using four input terminals for this purpose, which may be used with special test leads. Each of the two so-called Kelvin clips has a pair of jaws insulated from each other. One side of each clip applies the measuring current, while the other connections are only to sense the voltage drop. The resistance is again calculated using Ohm's Law as the measured voltage divided by the applied current.

Standards

Production Resistors

Resistor characteristics are quantified and reported using various national standards. In the US, MIL-STD-202 contains the relevant test methods to which other standards refer.

There are various standards specifying properties of resistors for use in equipment:

- BS 1852

- EIA-RS-279

- MIL-PRF-26

- MIL-PRF-39007 (Fixed Power, established reliability)

- MIL-PRF-55342 (Surface-mount thick and thin film)

- MIL-PRF-914

- MIL-R-11 STANDARD CANCELED

- MIL-R-39017 (Fixed, General Purpose, Established Reliability)

- MIL-PRF-32159 (zero ohm jumpers)

- UL 1412 (fusing and temperature limited resistors)

There are other United States military procurement MIL-R- standards.

Resistance Standards

The primary standard for resistance, the "mercury ohm" was initially defined in 1884 in as a column of mercury 106.3 cm long and 1 square millimeter in cross-section, at 0 degrees Celsius. Difficulties in precisely measuring the physical constants to replicate this standard result in variations of as much as 30 ppm. From 1900 the mercury ohm was replaced with a precision machined plate of manganin. Since 1990 the international resistance standard has been based on the quantized Hall effect discovered by Klaus von Klitzing, for which he won the Nobel Prize in Physics in 1985.

Resistors of extremely high precision are manufactured for calibration and laboratory use. They may have four terminals, using one pair to carry an operating current and the other pair to measure the voltage drop; this eliminates errors caused by voltage drops across the lead resistances, because no charge flows through voltage sensing leads. It is important in small value resistors (100–0.0001 ohm) where lead resistance is significant or even comparable with respect to resistance standard value.

Resistor Marking

Most axial resistors use a pattern of colored stripes to indicate resistance, which also indicate tolerance, and may also be extended to show temperature coefficient and reliability class. Cases are usually tan, brown, blue, or green, though other colors are occasionally found such as dark red or dark gray. The power rating is not usually marked and is deduced from the size.

The color bands of the carbon resistors can be three, four, five or, six bands. The first two bands represent first two digits to measure their value in ohms. The third band of a three- or four-banded resistor represents multiplier; a fourth band denotes tolerance (which if absent, denotes ±20%). For five and six color-banded resistors, the third band is a third digit, fourth band multiplier and fifth is tolerance. The sixth band represents temperature co-efficient in a six-banded resistor.

Surface-mount resistors are marked numerically, if they are big enough to permit marking; more-recent small sizes are impractical to mark.

Early 20th century resistors, essentially uninsulated, were dipped in paint to cover their entire body for color-coding. A second color of paint was applied to one end of the element, and a color dot (or band) in the middle provided the third digit. The rule was "body, tip, dot", providing two significant digits for value and the decimal multiplier, in that sequence. Default tolerance was ±20%. Closer-tolerance resistors had silver (±10%) or gold-colored (±5%) paint on the other end.

Preferred Values

Early resistors were made in more or less arbitrary round numbers; a series might have 100, 125, 150, 200, 300, etc. Resistors as manufactured are subject to a certain percentage tolerance, and it makes sense to manufacture values that correlate with the tolerance, so that the actual value of a resistor overlaps slightly with its neighbors. Wider spacing leaves gaps; narrower spacing increases manufacturing and inventory costs to provide resistors that are more or less interchangeable.

A logical scheme is to produce resistors in a range of values which increase in a geometric progression, so that each value is greater than its predecessor by a fixed multiplier or percentage, chosen to match the tolerance of the range. For example, for a tolerance of ±20% it makes sense to have each resistor about 1.5 times its predecessor, covering a decade in 6 values. In practice the factor used is 1.4678, giving values of 1.47, 2.15, 3.16, 4.64, 6.81, 10 for the 1–10-decade (a decade is a range increasing by a factor of 10; 0.1–1 and 10–100 are other examples); these are rounded in practice to 1.5, 2.2, 3.3, 4.7, 6.8, 10; followed, by 15, 22, 33, ... and preceded by ... 0.47, 0.68, 1. This scheme has been adopted as the E6 series of the IEC 60063 preferred number values. There are also E12, E24, E48, E96 and E192 series for components of progressively finer resolution, with 12, 24, 96, and 192 different values within each decade. The actual values used are in the IEC 60063 lists of preferred numbers.

A resistor of 100 ohms ±20% would be expected to have a value between 80 and 120 ohms; its E6 neighbors are 68 (54–82) and 150 (120–180) ohms. A sensible spacing, E6 is used for ±20% components; E12 for ±10%; E24 for ±5%; E48 for ±2%, E96 for ±1%; E192 for ±0.5% or better. Resistors are manufactured in values from a few milliohms to about a gigaohm in IEC60063 ranges appropriate for their tolerance. Manufacturers may sort resistors into tolerance-classes based on measurement. Accordingly, a selection of 100 ohms resistors with a tolerance of ±10%, might not lie just around 100 ohm (but no more than 10% off) as one would expect (a bell-curve), but rather be in two groups – either between 5 and 10% too high or 5 to 10% too low (but not closer to 100 ohm than that) because any resistors the factory had measured as being less than 5% off would have been marked and sold as resistors with only ±5% tolerance or better. When designing a circuit, this may become a consideration. This process of sorting parts based on post-production measurement is known as "binning", and can be applied to other components than resistors (such as speed grades for CPUs).

Earlier power wirewound resistors, such as brown vitreous-enameled types, however, were made with a different system of preferred values, such as some of those mentioned in the first sentence of this section.

SMT Resistors

This image shows four surface-mount resistors (the component at the upper left is a capacitor) including two zero-ohm resistors. Zero-ohm links are often used instead of wire links, so that they can be inserted by a resistor-inserting machine. Their resistance is non-zero but negligible.

Surface mounted resistors of larger sizes (metric 1608 and above) are printed with numerical values in a code related to that used on axial resistors. Standard-tolerance surface-mount technology (SMT) resistors are marked with a three-digit code, in which the first two digits are the first two significant digits of the value and the third digit is the power of ten (the number of zeroes). For example:

334	$= 33 \times 10^4$ ohms = 330 kilohms
222	$= 22 \times 10^2$ ohms = 2.2 kilohms
473	$= 47 \times 10^3$ ohms = 47 kilohms
105	$= 10 \times 10^5$ ohms = 1 megohm

Resistances less than 100 ohms are written: 100, 220, 470. The final zero represents ten to the power zero, which is 1. For example:

100	$= 10 \times 10^0$ ohm = 10 ohms
220	$= 22 \times 10^0$ ohm = 22 ohms

Sometimes these values are marked as *10* or *22* to prevent a mistake.

Resistances less than 10 ohms have 'R' to indicate the position of the decimal point (radix point). For example:

4R7	= 4.7 ohms
R300	= 0.30 ohms
0R22	= 0.22 ohms
0R01	= 0.01 ohms

Precision resistors are marked with a four-digit code, in which the first three digits are the significant figures and the fourth is the power of ten. For example:

1001	= 100 × 10^1 ohms = 1.00 kilohm
4992	= 499 × 10^2 ohms = 49.9 kilohm
1000	= 100 × 10^0 ohm = 100 ohms

000 and *0000* sometimes appear as values on surface-mount zero-ohm links, since these have (approximately) zero resistance.

More recent surface-mount resistors are too small, physically, to permit practical markings to be applied.

Industrial Type Designation

Format: *[two letters]<space>[resistance value (three digit)]<nospace>[tolerance code(numerical – one digit)]*

Power Rating at 70 °C			
Type No.	**Power rating (watts)**	**MIL-R-11 Style**	**MIL-R-39008 Style**
BB	⅛	RC05	RCR05
CB	¼	RC07	RCR07
EB	½	RC20	RCR20
GB	1	RC32	RCR32
HB	2	RC42	RCR42
GM	3	-	-
HM	4	-	-

Tolerance Code		
Industrial type designation	**Tolerance**	**MIL Designation**
5	±5%	J
2	±20%	M
1	±10%	K
-	±2%	G
-	±1%	F
-	±0.5%	D
-	±0.25%	C
-	±0.1%	B

Electrical and Thermal Noise

In amplifying faint signals, it is often necessary to minimize electronic noise, particularly in the first stage of amplification. As a dissipative element, even an ideal resistor naturally produces a randomly fluctuating voltage, or noise, across its terminals. This Johnson–Nyquist noise is a fundamental noise source which depends only upon the temperature and resistance of the resistor, and is predicted by the fluctuation–dissipation theorem. Using a larger value of resistance produces a larger voltage noise, whereas a smaller value of resistance generates more current noise, at a given temperature.

The thermal noise of a practical resistor may also be larger than the theoretical prediction and that increase is typically frequency-dependent. Excess noise of a practical resistor is observed only when current flows through it. This is specified in unit of μV/V/decade – μV of noise per volt applied across the resistor per decade of frequency. The μV/V/decade value is frequently given in dB so that a resistor with a noise index of 0 dB exhibits 1 μV (rms) of excess noise for each volt across the resistor in each frequency decade. Excess noise is thus an example of $1/f$ noise. Thick-film and carbon composition resistors generate more excess noise than other types at low frequencies. Wire-wound and thin-film resistors are often used for their better noise characteristics. Carbon composition resistors can exhibit a noise index of 0 dB while bulk metal foil resistors may have a noise index of –40 dB, usually making the excess noise of metal foil resistors insignificant. Thin film surface mount resistors typically have lower noise and better thermal stability than thick film surface mount resistors. Excess noise is also size-dependent: in general excess noise is reduced as the physical size of a resistor is increased (or multiple resistors are used in parallel), as the independently fluctuating resistances of smaller components tend to average out.

While not an example of "noise" per se, a resistor may act as a thermocouple, producing a small DC voltage differential across it due to the thermoelectric effect if its ends are at different temperatures. This induced DC voltage can degrade the precision of instrumentation amplifiers in particular. Such voltages appear in the junctions of the resistor leads with the circuit board and with the resistor body. Common metal film resistors show such an effect at a magnitude of about 20 μV/°C. Some carbon composition resistors can exhibit thermoelectric offsets as high as 400 μV/°C, whereas specially constructed resistors can reduce this number to 0.05 μV/°C. In applications where the thermoelectric effect may become important, care has to be taken to mount the resistors horizontally to avoid temperature gradients and to mind the air flow over the board.

Failure Modes

The failure rate of resistors in a properly designed circuit is low compared to other electronic components such as semiconductors and electrolytic capacitors. Damage to resistors most often occurs due to overheating when the average power delivered to it greatly exceeds its ability to dissipate heat (specified by the resistor's *power rating*). This may be due to a fault external to the circuit, but is frequently caused by the failure of another component (such as a transistor that shorts out) in the circuit connected to the resistor. Operating a resistor too close to its power rating can limit the resistor's lifespan or cause a significant change in its resistance. A safe design generally uses overrated resistors in power applications to avoid this danger.

Low-power thin-film resistors can be damaged by long-term high-voltage stress, even below maximum specified voltage and below maximum power rating. This is often the case for the startup resistors feeding the SMPS integrated circuit.

When overheated, carbon-film resistors may decrease or increase in resistance. Carbon film and composition resistors can fail (open circuit) if running close to their maximum dissipation. This is also possible but less likely with metal film and wirewound resistors.

There can also be failure of resistors due to mechanical stress and adverse environmental factors including humidity. If not enclosed, wirewound resistors can corrode.

Surface mount resistors have been known to fail due to the ingress of sulfur into the internal makeup of the resistor. This sulfur chemically reacts with the silver layer to produce non-conductive silver sulfide. The resistor's impedance goes to infinity. Sulfur resistant and anti-corrosive resistors are sold into automotive, industrial, and military applications. ASTM B809 is an industry standard that tests a part's susceptibility to sulfur.

An alternative failure mode can be encountered where large value resistors are used (hundreds of kilohms and higher). Resistors are not only specified with a maximum power dissipation, but also for a maximum voltage drop. Exceeding this voltage causes the resistor to degrade slowly reducing in resistance. The voltage dropped across large value resistors can be exceeded before the power dissipation reaches its limiting value. Since the maximum voltage specified for commonly encountered resistors is a few hundred volts, this is a problem only in applications where these voltages are encountered.

Variable resistors can also degrade in a different manner, typically involving poor contact between the wiper and the body of the resistance. This may be due to dirt or corrosion and is typically perceived as "crackling" as the contact resistance fluctuates; this is especially noticed as the device is adjusted. This is similar to crackling caused by poor contact in switches, and like switches, potentiometers are to some extent self-cleaning: running the wiper across the resistance may improve the contact. Potentiometers which are seldom adjusted, especially in dirty or harsh environments, are most likely to develop this problem. When self-cleaning of the contact is insufficient, improvement can usually be obtained through the use of contact cleaner (also known as "tuner cleaner") spray. The crackling noise associated with turning the shaft of a dirty potentiometer in an audio circuit (such as the volume control) is greatly accentuated when an undesired DC voltage is present, often indicating the failure of a DC blocking capacitor in the circuit.

Transistor

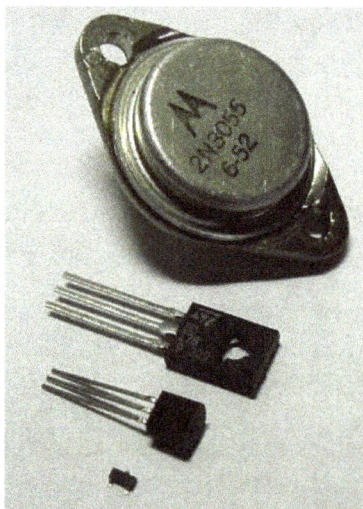

Assorted discrete transistors. Packages in order from top to bottom: TO-3, TO-126, TO-92, SOT-23.

A transistor is a semiconductor device used to amplify or switch electronic signals and electrical power. It is composed of semiconductor material usually with at least three terminals for connection to an external circuit. A voltage or current applied to one pair of the transistor's terminals controls the current through another pair of terminals. Because the controlled (output) power can be higher than the controlling (input) power, a transistor can amplify a signal. Today, some transistors are packaged individually, but many more are found embedded in integrated circuits.

The transistor is the fundamental building block of modern electronic devices, and is ubiquitous in modern electronic systems. Julius Lilienfeld patented a field-effect transistor in 1926 but it was not possible to actually construct a working device at that time. The first practically implemented device was a point-contact transistor invented in 1947 by American physicists John Bardeen, Walter Brattain, and William Shockley. The transistor revolutionized the field of electronics, and paved the way for smaller and cheaper radios, calculators, and computers, among other things. The transistor is on the list of IEEE milestones in electronics, and Bardeen, Brattain, and Shockley shared the 1956 Nobel Prize in Physics for their achievement.

History

The thermionic triode, a vacuum tube invented in 1907, enabled amplified radio technology and long-distance telephony. The triode, however, was a fragile device that consumed a lot of power. Physicist Julius Edgar Lilienfeld filed a patent for a field-effect transistor (FET) in Canada in 1925, which was intended to be a solid-state replacement for the triode. Lilienfeld also filed identical patents in the United States in 1926 and 1928. However, Lilienfeld did not publish any research articles about his devices nor did his patents cite any specific examples of a working prototype. Because the production of high-quality semiconductor materials was still decades away, Lilienfeld's solid-state amplifier ideas would not have found practical use in the 1920s and 1930s, even if such a device had been built. In 1934, German inventor Oskar Heil patented a similar device in Europe.

John Bardeen, William Shockley and Walter Brattain at Bell Labs, 1948.

From November 17, 1947 to December 23, 1947, John Bardeen and Walter Brattain at AT&T's Bell Labs in the United States performed experiments and observed that when two gold point contacts were applied to a crystal of germanium, a signal was produced with the output power greater than the input. Solid State Physics Group leader William Shockley saw the potential in this, and over the next few months worked to greatly expand the knowledge of semiconductors. The term *transistor* was coined by John R. Pierce as a contraction of the term *transresistance*. According to Lillian Hoddeson and Vicki Daitch, authors of a biography of John Bardeen, Shockley had proposed that Bell Labs' first patent for a transistor should be based on the field-effect and that he be named as the inventor. Having unearthed Lilienfeld's patents that went into obscurity years earlier, lawyers at Bell Labs advised against Shockley's proposal because the idea of a field-effect transistor that used an electric field as a "grid" was not new. Instead, what Bardeen, Brattain, and Shockley invented in 1947 was the first point-contact transistor. In acknowledgement of this accomplishment, Shockley, Bardeen, and Brattain were jointly awarded the 1956 Nobel Prize in Physics "for their researches on semiconductors and their discovery of the transistor effect."

Herbert F. Mataré (1950)

In 1948, the point-contact transistor was independently invented by German physicists Herbert Mataré and Heinrich Welker while working at the Compagnie des Freins et Signaux, a Westinghouse subsidiary located in Paris. Mataré had previous experience in developing crystal rectifiers from silicon and germanium in the German radar effort during World War II. Using this knowledge, he began researching the phenomenon of "interference" in 1947. By June 1948, witnessing currents flowing through point-contacts, Mataré produced consistent results using samples of germanium produced by Welker, similar to what Bardeen and Brattain had accomplished earlier in December 1947. Realizing that Bell Labs' scientists had already invented the transistor before them, the company rushed to get its "transistron" into production for amplified use in France's telephone network.

The first high-frequency transistor was the surface-barrier germanium transistor developed by Philco in 1953, capable of operating up to 60 MHz. These were made by etching depressions into an N-type germanium base from both sides with jets of Indium(III) sulfate until it was a few ten-thousandths of an inch thick. Indium electroplated into the depressions formed the collector and emitter.

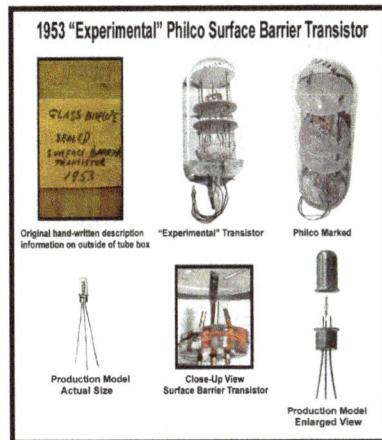

Philco surface-barrier transistor developed and produced in 1953

The first "prototype" pocket transistor radio was shown by INTERMETALL (a company founded by Herbert Mataré in 1952) at the Internationale Funkausstellung Düsseldorf between August 29, 1953 and September 9, 1953.

The first "production" all-transistor car radio was produced in 1955 by Chrysler and Philco, had used surface-barrier transistors in its circuitry and which were also first suitable for high-speed computers.

The first working silicon transistor was developed at Bell Labs on January 26, 1954 by Morris Tanenbaum. The first commercial silicon transistor was produced by Texas Instruments in 1954. This was the work of Gordon Teal, an expert in growing crystals of high purity, who had previously worked at Bell Labs. The first MOS transistor actually built was by Kahng and Atalla at Bell Labs in 1960.

Importance

A Darlington transistor opened up so the actual transistor chip (the small square) can be seen inside. A Darlington transistor is effectively two transistors on the same chip. One transistor is much larger than the other, but both are large in comparison to transistors in large-scale integration because this particular example is intended for power applications.

The transistor is the key active component in practically all modern electronics. Many consider it to be one of the greatest inventions of the 20th century. Its importance in today's society rests on its ability to be mass-produced using a highly automated process (semiconductor device fabrication) that achieves astonishingly low per-transistor costs. The invention of the first transistor at Bell Labs was named an IEEE Milestone in 2009.

Although several companies each produce over a billion individually packaged (known as *discrete*) transistors every year, the vast majority of transistors are now produced in integrated circuits (often shortened to *IC*, *microchips* or simply *chips*), along with diodes, resistors, capacitors and other electronic components, to produce complete electronic circuits. A logic gate consists of up to about twenty transistors whereas an advanced microprocessor, as of 2009, can use as many as 3 billion transistors (MOSFETs). "About 60 million transistors were built in 2002... for [each] man, woman, and child on Earth."

The transistor's low cost, flexibility, and reliability have made it a ubiquitous device. Transistorized mechatronic circuits have replaced electromechanical devices in controlling appliances and machinery. It is often easier and cheaper to use a standard microcontroller and write a computer program to carry out a control function than to design an equivalent mechanical control function.

Simplified Operation

The essential usefulness of a transistor comes from its ability to use a small signal applied between one pair of its terminals to control a much larger signal at another pair of terminals. This property is called gain. It can produce a stronger output signal, a voltage or current, which is proportional to a weaker input signal; that is, it can act as an amplifier. Alternatively, the transistor can be used to turn current on or off in a circuit as an electrically controlled switch, where the amount of current is determined by other circuit elements.

There are two types of transistors, which have slight differences in how they are used in a circuit. A *bipolar transistor* has terminals labeled base, collector, and emitter. A small current at the base terminal (that is, flowing between the base and the emitter) can control or switch a much larger current between the collector and emitter terminals. For a *field-effect transistor*, the terminals are labeled gate, source, and drain, and a voltage at the gate can control a current between source and drain.

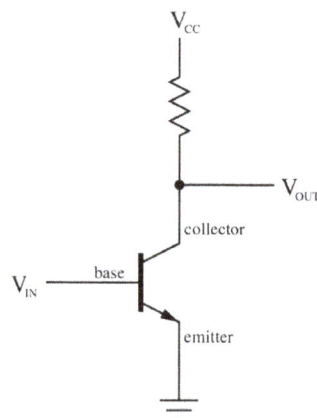

A simple circuit diagram to show the labels of a n–p–n bipolar transistor.

The image represents a typical bipolar transistor in a circuit. Charge will flow between emitter and collector terminals depending on the current in the base. Because internally the base and emitter connections behave like a semiconductor diode, a voltage drop develops between base and emitter while the base current exists. The amount of this voltage depends on the material the transistor is made from, and is referred to as V_{BE}.

Transistor as a Switch

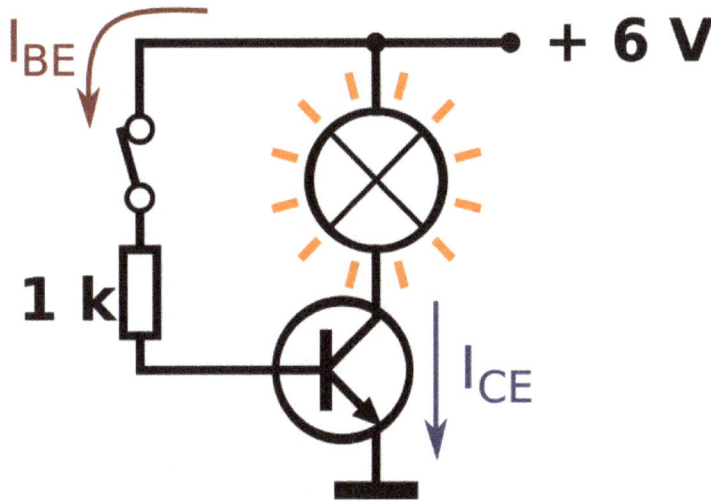

BJT used as an electronic switch, in grounded-emitter configuration.

Transistors are commonly used in digital circuits as electronic switches which can be either in an "on" or "off" state, both for high-power applications such as switched-mode power supplies and for low-power applications such as logic gates. Important parameters for this application include the current switched, the voltage handled, and the switching speed, characterised by the rise and fall times.

In a grounded-emitter transistor circuit, such as the light-switch circuit shown, as the base voltage rises, the emitter and collector currents rise exponentially. The collector voltage drops because of reduced resistance from collector to emitter. If the voltage difference between the collector and emitter were zero (or near zero), the collector current would be limited only by the load resistance (light bulb) and the supply voltage. This is called *saturation* because current is flowing from collector to emitter freely. When saturated, the switch is said to be *on*.

Providing sufficient base drive current is a key problem in the use of bipolar transistors as switches. The transistor provides current gain, allowing a relatively large current in the collector to be switched by a much smaller current into the base terminal. The ratio of these currents varies depending on the type of transistor, and even for a particular type, varies depending on the collector current. In the example light-switch circuit shown, the resistor is chosen to provide enough base current to ensure the transistor will be saturated.

In a switching circuit, the idea is to simulate, as near as possible, the ideal switch having the properties of open circuit when off, short circuit when on, and an instantaneous transition between the two states. Parameters are chosen such that the "off" output is limited to leakage currents too small to affect connected circuitry; the resistance of the transistor in the "on" state is too small to affect circuitry; and the transition between the two states is fast enough not to have a detrimental effect.

Transistor as an Amplifier

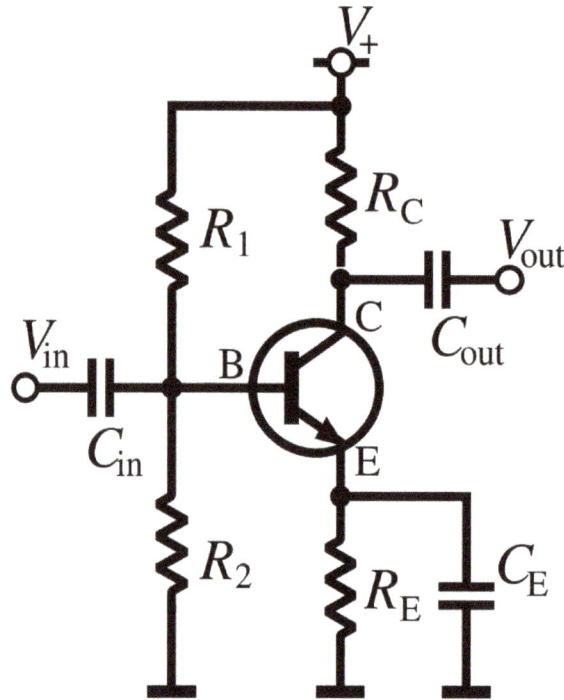

Amplifier circuit, common-emitter configuration with a voltage-divider bias circuit.

The common-emitter amplifier is designed so that a small change in voltage (V_{in}) changes the small current through the base of the transistor; the transistor's current amplification combined with the properties of the circuit mean that small swings in V_{in} produce large changes in V_{out}.

Various configurations of single transistor amplifier are possible, with some providing current gain, some voltage gain, and some both.

From mobile phones to televisions, vast numbers of products include amplifiers for sound reproduction, radio transmission, and signal processing. The first discrete-transistor audio amplifiers barely supplied a few hundred milliwatts, but power and audio fidelity gradually increased as better transistors became available and amplifier architecture evolved.

Modern transistor audio amplifiers of up to a few hundred watts are common and relatively inexpensive.

Comparison with Vacuum Tubes

Before transistors were developed, vacuum (electron) tubes (or in the UK "thermionic valves" or just "valves") were the main active components in electronic equipment.

Advantages

The key advantages that have allowed transistors to replace vacuum tubes in most applications are

- no cathode heater (which produces the characteristic orange glow of tubes), reducing power consumption, eliminating delay as tube heaters warm up, and immune from cathode poisoning and depletion;

- very small size and weight, reducing equipment size;

- large numbers of extremely small transistors can be manufactured as a single integrated circuit;

- low operating voltages compatible with batteries of only a few cells;

- circuits with greater energy efficiency are usually possible. For low-power applications (e.g., voltage amplification) in particular, energy consumption can be very much less than for tubes;

- inherent reliability and very long life; tubes always degrade and fail over time. Some transistorized devices have been in service for more than 50 years ;

- complementary devices available, providing design flexibility including complementary-symmetry circuits, not possible with vacuum tubes;

- very low sensitivity to mechanical shock and vibration, providing physical ruggedness and virtually eliminating shock-induced spurious signals (e.g., microphonics in audio applications);

- not susceptible to breakage of a glass envelope, leakage, outgassing, and other physical damage.

Limitations

Transistors have the following limitations:

- silicon transistors can age and fail;

- high-power, high-frequency operation, such as that used in over-the-air television broadcasting, is better achieved in vacuum tubes due to improved electron mobility in a vacuum;

- solid-state devices are susceptible to damage from very brief electrical and thermal events, including electrostatic discharge in handling; vacuum tubes are electrically much more rugged;

- sensitivity to radiation and cosmic rays (special radiation-hardened chips are used for spacecraft devices);

- vacuum tubes in audio applications create significant lower-harmonic distortion, the so-called tube sound, which some people prefer.

Types

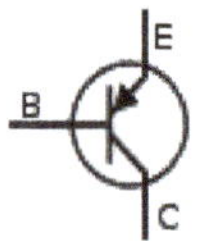	PNP	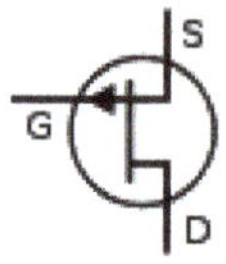	P-channel
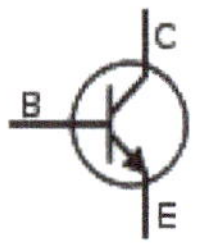	NPN	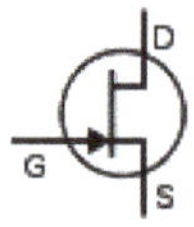	N-channel
BJT		JFET	

BJT and JFET symbols

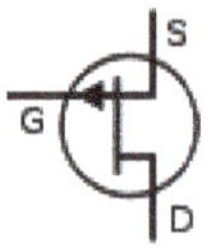	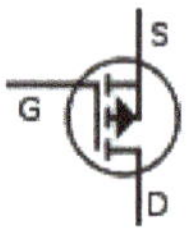	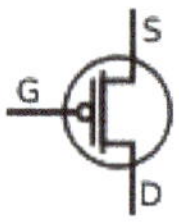	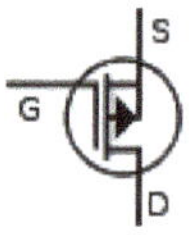	P-channel
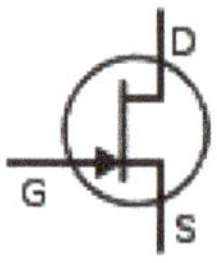	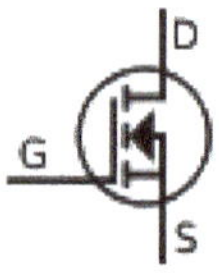	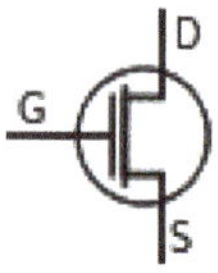	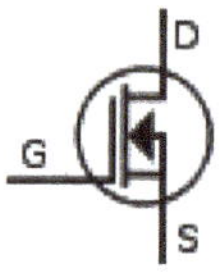	N-channel
JFET	MOSFET enh		MOSFET dep	

JFET and MOSFET symbols

Transistors are categorized by

- semiconductor material: the metalloids germanium (first used in 1947) and silicon (first used in 1954)—in amorphous, polycrystalline and monocrystalline form—, the compounds gallium arsenide (1966) and silicon carbide (1997), the alloy silicon-germanium (1989), the allotrope of carbon graphene (research ongoing since 2004), etc.

- structure: BJT, JFET, IGFET (MOSFET), insulated-gate bipolar transistor, "other types";

- electrical polarity (positive and negative): n–p–n, p–n–p (BJTs), n-channel, p-channel (FETs);

- maximum power rating: low, medium, high;

- maximum operating frequency: low, medium, high, radio (RF), microwave frequency (the maximum effective frequency of a transistor in a common-emitter or common-source circuit is denoted by the term f_T, an abbreviation for transition frequency—the frequency of transition is the frequency at which the transistor yields unity voltage gain)

- application: switch, general purpose, audio, high voltage, super-beta, matched pair;

- physical packaging: through-hole metal, through-hole plastic, surface mount, ball grid array, power modules;

- amplification factor h_{FE}, β_F (transistor beta) or g_m (transconductance).

Hence, a particular transistor may be described as *silicon, surface-mount, BJT, n–p–n, low-power, high-frequency switch.*

A popular way to remember which symbol represents which type of transistor is to look at the arrow and how it is arranged. Within an NPN transistor symbol, the arrow will Not Point iN. Conversely, within the PNP symbol you see that the arrow Points iN Proudly.

Bipolar Junction Transistor (Bjt)

Bipolar transistors are so named because they conduct by using both majority and minority carriers. The bipolar junction transistor, the first type of transistor to be mass-produced, is a combination of two junction diodes, and is formed of either a thin layer of p-type semiconductor sandwiched between two n-type semiconductors (an n–p–n transistor), or a thin layer of n-type semiconductor sandwiched between two p-type semiconductors (a p–n–p transistor). This construction produces two p–n junctions: a base–emitter junction and a base–collector junction, separated by a thin region of semiconductor known as the base region (two junction diodes wired together without sharing an intervening semiconducting region will not make a transistor).

BJTs have three terminals, corresponding to the three layers of semiconductor—an *emitter*, a *base*, and a *collector*. They are useful in amplifiers because the currents at the emitter and collector are controllable by a relatively small base current. In an n–p–n transistor operating in the active region, the emitter–base junction is forward biased (electrons and holes recombine at the junction), and electrons are injected into the base region. Because the base is narrow, most of these electrons will diffuse into the reverse-biased (electrons and holes are formed at, and move away from the junction) base–collector junction and be swept into the collector; perhaps one-hundredth of the electrons will recombine in the base, which is the dominant mechanism in the base current. By controlling the number of electrons that can leave the base, the number of electrons entering the collector can be controlled. Collector current is approximately β (common-emitter current gain) times the base current. It is typically greater than 100 for small-signal transistors but can be smaller in transistors designed for high-power applications.

Unlike the field-effect transistor (see below), the BJT is a low-input-impedance device. Also, as the base–emitter voltage (V_{BE}) is increased the base–emitter current and hence the collector–emitter current (I_{CE}) increase exponentially according to the Shockley diode model and the Ebers-Moll model. Because of this exponential relationship, the BJT has a higher transconductance than the FET.

Bipolar transistors can be made to conduct by exposure to light, because absorption of photons in the base region generates a photocurrent that acts as a base current; the collector current is approximately β times the photocurrent. Devices designed for this purpose have a transparent window in the package and are called phototransistors.

Field-effect Transistor (FET)

Operation of a FET and its Id-Vg curve. At first, when no gate voltage is applied. There is no inversion electron in the channel, the device is OFF. As gate voltage increase, inversion electron density in the channel increase, current increase, the device turns on.

The *field-effect transistor*, sometimes called a *unipolar transistor*, uses either electrons (in *n-channel FET*) or holes (in *p-channel FET*) for conduction. The four terminals of the FET are named *source, gate, drain,* and *body* (*substrate*). On most FETs, the body is connected to the source inside the package, and this will be assumed for the following description.

In a FET, the drain-to-source current flows via a conducting channel that connects the *source* region to the *drain* region. The conductivity is varied by the electric field that is produced when a voltage is applied between the gate and source terminals; hence the current flowing between the drain and source is controlled by the voltage applied between the gate and source. As the gate–source voltage (V_{GS}) is increased, the drain–source current (I_{DS}) increases exponentially for V_{GS} below threshold, and then at a roughly quadratic rate ($I_{GS} \propto (V_{GS} - V_{T})^2$) (where V_{T} is the threshold voltage at which drain current begins) in the "space-charge-limited" region above threshold. A quadratic behavior is not observed in modern devices, for example, at the 65 nm technology node.

For low noise at narrow bandwidth the higher input resistance of the FET is advantageous.

FETs are divided into two families: *junction FET* (JFET) and *insulated gate FET* (IGFET). The IGFET is more commonly known as a *metal–oxide–semiconductor FET* (MOSFET), reflecting its original construction from layers of metal (the gate), oxide (the insulation), and semiconductor. Unlike IGFETs, the JFET gate forms a p–n diode with the channel which lies between the source and drain. Functionally, this makes the n-channel JFET the solid-state equivalent of the vacuum tube triode which, similarly, forms a diode between its grid and cathode. Also, both devices operate in the *depletion mode*, they both have a high input impedance, and they both conduct current under the control of an input voltage.

Metal–semiconductor FETs (MESFETs) are JFETs in which the reverse biased p–n junction is replaced by a metal–semiconductor junction. These, and the HEMTs (high-electron-mobility transistors, or HFETs), in which a two-dimensional electron gas with very high carrier mobility is used for charge transport, are especially suitable for use at very high frequencies (microwave frequencies; several GHz).

FETs are further divided into *depletion-mode* and *enhancement-mode* types, depending on whether the channel is turned on or off with zero gate-to-source voltage. For enhancement mode, the channel is off at zero bias, and a gate potential can "enhance" the conduction. For the depletion mode, the channel is on at zero bias, and a gate potential (of the opposite polarity) can "deplete" the channel, reducing conduction. For either mode, a more positive gate voltage corresponds to a higher current for n-channel devices and a lower current for p-channel devices. Nearly all JFETs are depletion-mode because the diode junctions would forward bias and conduct if they were enhancement-mode devices; most IGFETs are enhancement-mode types.

Usage of Bipolar and Field-effect Transistors

The bipolar junction transistor (BJT) was the most commonly used transistor in the 1960s and 70s. Even after MOSFETs became widely available, the BJT remained the transistor of choice for many analog circuits such as amplifiers because of their greater linearity and ease of manufacture. In integrated circuits, the desirable properties of MOSFETs allowed them to capture nearly all market share for digital circuits. Discrete MOSFETs can be applied in transistor applications, including analog circuits, voltage regulators, amplifiers, power transmitters and motor drivers.

Other Transistor Types

Transistor symbol created on Portuguese pavement in the University of Aveiro.

- Bipolar junction transistor (BJT):

    o   heterojunction bipolar transistor, up to several hundred GHz, common in modern ultrafast and RF circuits;

    o   Schottky transistor;

    o   avalanche transistor:

    o   Darlington transistors are two BJTs connected together to provide a high current gain equal to the product of the current gains of the two transistors;

    o   insulated-gate bipolar transistors (IGBTs) use a medium-power IGFET, similarly connected to a power BJT, to give a high input impedance. Power diodes are often connected between certain terminals depending on specific use. IGBTs are particularly suitable for heavy-duty industrial applications. The ASEA Brown Boveri (ABB) 5SNA2400E170100 illustrates just how far power semiconductor technology has advanced. Intended for three-phase power supplies, this device houses three n–p–n IGBTs in a case measuring 38 by 140 by 190 mm and weighing 1.5 kg. Each IGBT is rated at 1,700 volts and can handle 2,400 amperes;

    o   phototransistor;

- o multiple-emitter transistor, used in transistor–transistor logic and integrated current mirrors;

- o multiple-base transistor, used to amplify very-low-level signals in noisy environments such as the pickup of a record player or radio front ends. Effectively, it is a very large number of transistors in parallel where, at the output, the signal is added constructively, but random noise is added only stochastically.

- Field-effect transistor (FET):

  - o carbon nanotube field-effect transistor (CNFET), where the channel material is replaced by a carbon nanotube;

  - o junction gate field-effect transistor (JFET), where the gate is insulated by a reverse-biased p–n junction;

  - o metal–semiconductor field-effect transistor (MESFET), similar to JFET with a Schottky junction instead of a p–n junction;

    - ☐ high-electron-mobility transistor (HEMT);

  - o metal–oxide–semiconductor field-effect transistor (MOSFET), where the gate is insulated by a shallow layer of insulator;

  - o inverted-T field-effect transistor (ITFET);

  - o fin field-effect transistor (FinFET), source/drain region shapes fins on the silicon surface;

  - o fast-reverse epitaxial diode field-effect transistor (FREDFET);

  - o thin-film transistor, in LCDs;

  - o organic field-effect transistor (OFET), in which the semiconductor is an organic compound;

  - o ballistic transistor;

  - o floating-gate transistor, for non-volatile storage;

  - o FETs used to sense environment;

    - ☐ ion-sensitive field-effect transistor (IFSET), to measure ion concentrations in solution,

    - ☐ electrolyte–oxide–semiconductor field-effect transistor (EOSFET), neurochip,

    - ☐ deoxyribonucleic acid field-effect transistor (DNAFET).

- Tunnel field-effect transistor, where it switches by modulating quantum tunnelling through a barrier.

- Diffusion transistor, formed by diffusing dopants into semiconductor substrate; can be both BJT and FET.

- Unijunction transistor, can be used as simple pulse generators. It comprise a main body of either P-type or N-type semiconductor with ohmic contacts at each end (terminals *Base1* and *Base2*). A junction with the opposite semiconductor type is formed at a point along the length of the body for the third terminal (*Emitter*).

- Single-electron transistors (SET), consist of a gate island between two tunneling junctions. The tunneling current is controlled by a voltage applied to the gate through a capacitor.

- Nanofluidic transistor, controls the movement of ions through sub-microscopic, water-filled channels.

- Multigate devices:

  o   tetrode transistor;

  o   pentode transistor;

  o   trigate transistor (prototype by Intel);

  o   dual-gate field-effect transistors have a single channel with two gates in cascode; a configuration optimized for *high-frequency amplifiers*, *mixers*, and oscillators.

- Junctionless nanowire transistor (JNT), uses a simple nanowire of silicon surrounded by an electrically isolated "wedding ring" that acts to gate the flow of electrons through the wire.

- Vacuum-channel transistor, when in 2012, NASA and the National Nanofab Center in South Korea were reported to have built a prototype vacuum-channel transistor in only 150 nanometers in size, can be manufactured cheaply using standard silicon semiconductor processing, can operate at high speeds even in hostile environments, and could consume just as much power as a standard transistor.

- Organic electrochemical transistor.

Part Numbering Standards/Specifications

The types of some transistors can be parsed from the part number. There are three major semiconductor naming standards; in each the alphanumeric prefix provides clues to type of the device.

Japanese Industrial Standard (JIS)

JIS Transistor Prefix Table	
Prefix	**Type of transistor**
2SA	high-frequency p–n–p BJTs
2SB	audio-frequency p–n–p BJTs

2SC	high-frequency n–p–n BJTs
2SD	audio-frequency n–p–n BJTs
2SJ	P-channel FETs (both JFETs and MOSFETs)
2SK	N-channel FETs (both JFETs and MOSFETs)

The *JIS-C-7012* specification for transistor part numbers starts with "2S", e.g. 2SD965, but sometimes the "2S" prefix is not marked on the package – a 2SD965 might only be marked "D965"; a 2SC1815 might be listed by a supplier as simply "C1815". This series sometimes has suffixes (such as "R", "O", "BL", standing for "red", "orange", "blue", etc.) to denote variants, such as tighter h_{FE} (gain) groupings.

European Electronic Component Manufacturers Association (EECA)

The Pro Electron standard, the European Electronic Component Manufacturers Association part numbering scheme, begins with two letters: the first gives the semiconductor type (A for germanium, B for silicon, and C for materials like GaAs); the second letter denotes the intended use (A for diode, C for general-purpose transistor, etc.). A 3-digit sequence number (or one letter then 2 digits, for industrial types) follows. With early devices this indicated the case type. Suffixes may be used, with a letter (e.g. "C" often means high h_{FE}, such as in: BC549C) or other codes may follow to show gain (e.g. BC327-25) or voltage rating (e.g. BUK854-800A). The more common prefixes are:

Pro Electron / EECA Transistor Prefix Table					
Prefix class	**Type and usage**	**Example**	**Equivalent**	**Reference**	
AC	Germanium small-signal AF transistor	AC126	NTE102A	Datasheet	
AD	Germanium AF power transistor	AD133	NTE179	Datasheet	
AF	Germanium small-signal RF transistor	AF117	NTE160	Datasheet	
AL	Germanium RF power transistor	ALZ10	NTE100	Datasheet	
AS	Germanium switching transistor	ASY28	NTE101	Datasheet	
AU	Germanium power switching transistor	AU103	NTE127	Datasheet	
BC	Silicon, small-signal transistor ("general purpose")	BC548	2N3904	Datasheet	
BD	Silicon, power transistor	BD139	NTE375	Datasheet	
BF	Silicon, RF (high frequency) BJT or FET	BF245	NTE133	Datasheet	
BS	Silicon, switching transistor (BJT or MOSFET)	BS170	2N7000	Datasheet	
BL	Silicon, high frequency, high power (for transmitters)	BLW60	NTE325	Datasheet	
BU	Silicon, high voltage (for CRT horizontal deflection circuits)	BU2520A	NTE2354	Datasheet	
CF	Gallium Arsenide small-signal Microwave transistor (MESFET)	CF739	–	Datasheet	
CL	Gallium Arsenide Microwave power transistor (FET)	CLY10	–	Datasheet	

Joint Electron Devices Engineering Council (JEDEC)

The JEDEC *EIA370* transistor device numbers usually start with "2N", indicating a three-terminal device (dual-gate field-effect transistors are four-terminal devices, so begin with 3N), then a 2, 3 or 4-digit sequential number with no significance as to device properties (although early devices with low numbers tend to be germanium). For example, 2N3055 is a silicon n–p–n power transistor, 2N1301 is a p–n–p germanium switching transistor. A letter suffix (such as "A") is sometimes used to indicate a newer variant, but rarely gain groupings.

Proprietary

Manufacturers of devices may have their own proprietary numbering system, for example CK722. Since devices are second-sourced, a manufacturer's prefix (like "MPF" in MPF102, which originally would denote a Motorola FET) now is an unreliable indicator of who made the device. Some proprietary naming schemes adopt parts of other naming schemes, for example a PN2222A is a (possibly Fairchild Semiconductor) 2N2222A in a plastic case (but a PN108 is a plastic version of a BC108, not a 2N108, while the PN100 is unrelated to other xx100 devices).

Military part numbers sometimes are assigned their own codes, such as the British Military CV Naming System.

Manufacturers buying large numbers of similar parts may have them supplied with "house numbers", identifying a particular purchasing specification and not necessarily a device with a standardized registered number. For example, an HP part 1854,0053 is a (JEDEC) 2N2218 transistor which is also assigned the CV number: CV7763

Naming Problems

With so many independent naming schemes, and the abbreviation of part numbers when printed on the devices, ambiguity sometimes occurs. For example, two different devices may be marked "J176" (one the J176 low-power JFET, the other the higher-powered MOSFET 2SJ176).

As older "through-hole" transistors are given surface-mount packaged counterparts, they tend to be assigned many different part numbers because manufacturers have their own systems to cope with the variety in pinout arrangements and options for dual or matched n–p–n+p–n–p devices in one pack. So even when the original device (such as a 2N3904) may have been assigned by a standards authority, and well known by engineers over the years, the new versions are far from standardized in their naming.

Construction

Semiconductor Material

Semiconductor material characteristics				
Semiconductor material	Junction forward voltage V @ 25 °C	Electron mobility m²/(V·s) @ 25 °C	Hole mobility m²/(V·s) @ 25 °C	Max. junction temp. °C

Ge	0.27	0.39	0.19	70 to 100
Si	0.71	0.14	0.05	150 to 200
GaAs	1.03	0.85	0.05	150 to 200
Al-Si junction	0.3	—	—	150 to 200

The first BJTs were made from germanium (Ge). Silicon (Si) types currently predominate but certain advanced microwave and high-performance versions now employ the *compound semiconductor* material gallium arsenide (GaAs) and the *semiconductor alloy* silicon germanium (SiGe). Single element semiconductor material (Ge and Si) is described as *elemental.*

Rough parameters for the most common semiconductor materials used to make transistors are given in the adjacent table; these parameters will vary with increase in temperature, electric field, impurity level, strain, and sundry other factors.

The *junction forward voltage* is the voltage applied to the emitter–base junction of a BJT in order to make the base conduct a specified current. The current increases exponentially as the junction forward voltage is increased. The values given in the table are typical for a current of 1 mA (the same values apply to semiconductor diodes). The lower the junction forward voltage the better, as this means that less power is required to "drive" the transistor. The junction forward voltage for a given current decreases with increase in temperature. For a typical silicon junction the change is −2.1 mV/°C. In some circuits special compensating elements (sensistors) must be used to compensate for such changes.

The density of mobile carriers in the channel of a MOSFET is a function of the electric field forming the channel and of various other phenomena such as the impurity level in the channel. Some impurities, called dopants, are introduced deliberately in making a MOSFET, to control the MOSFET electrical behavior.

The *electron mobility* and *hole mobility* columns show the average speed that electrons and holes diffuse through the semiconductor material with an electric field of 1 volt per meter applied across the material. In general, the higher the electron mobility the faster the transistor can operate. The table indicates that Ge is a better material than Si in this respect. However, Ge has four major shortcomings compared to silicon and gallium arsenide:

- Its maximum temperature is limited;
- it has relatively high leakage current;
- it cannot withstand high voltages;
- it is less suitable for fabricating integrated circuits.

Because the electron mobility is higher than the hole mobility for all semiconductor materials, a given bipolar n–p–n transistor tends to be swifter than an equivalent p–n–p transistor. GaAs has the highest electron mobility of the three semiconductors. It is for this reason that GaAs is used in high-frequency applications. A relatively recent FET development, the *high-electron-mobility transistor* (HEMT), has a heterostructure (junction between different semiconductor materials) of aluminium gallium arsenide (AlGaAs)-gallium arsenide (GaAs) which has twice the electron mobility of a GaAs-metal barrier junction. Because of their high speed and low noise, HEMTs are used in satellite receivers working at frequencies around 12 GHz. HEMTs based on gallium nitride and aluminium gallium nitride (AlGaN/GaN HEMTs) provide a still higher electron mobility and are being developed for various applications.

Max. junction temperature values represent a cross section taken from various manufacturers' data sheets. This temperature should not be exceeded or the transistor may be damaged.

Al–Si junction refers to the high-speed (aluminum–silicon) metal–semiconductor barrier diode, commonly known as a Schottky diode. This is included in the table because some silicon power IGFETs have a *parasitic* reverse Schottky diode formed between the source and drain as part of the fabrication process. This diode can be a nuisance, but sometimes it is used in the circuit.

Packaging

Assorted discrete transistors.

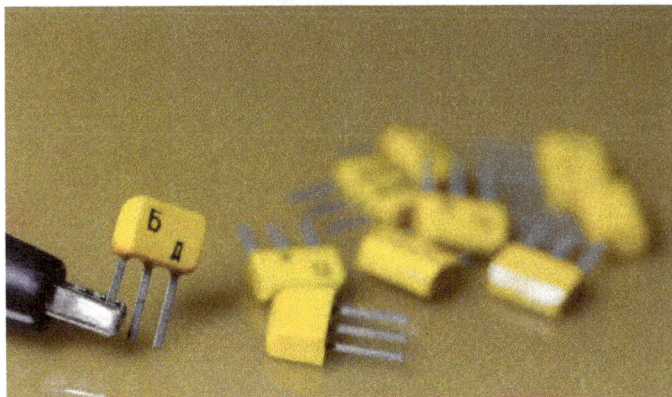

Soviet KT315b transistors.

Discrete transistors are individually packaged transistors. Transistors come in many different semiconductor packages. The two main categories are *through-hole* (or *leaded*), and *surface-mount*, also known as *surface-mount device* (SMD). The *ball grid array* (BGA) is the latest surface-mount package (currently only for large integrated circuits). It has solder "balls" on the underside in place of leads. Because they are smaller and have shorter interconnections, SMDs have better high-frequency characteristics but lower power rating.

Transistor packages are made of glass, metal, ceramic, or plastic. The package often dictates the power rating and frequency characteristics. Power transistors have larger packages that can be clamped to heat sinks for enhanced cooling. Additionally, most power transistors have the collector or drain physically connected to the metal enclosure. At the other extreme, some surface-mount *microwave* transistors are as small as grains of sand.

Often a given transistor type is available in several packages. Transistor packages are mainly standardized, but the assignment of a transistor's functions to the terminals is not: other transistor types can assign other functions to the package's terminals. Even for the same transistor type the terminal assignment can vary (normally indicated by a suffix letter to the part number, q.e. BC212L and BC212K).

Nowadays most transistors come in a wide range of SMT packages, in comparison the list of available through-hole packages is relatively small, here is a short list of the most common through-hole transistors packages in alphabetical order: ATV, E-line, MRT, HRT, SC-43, SC-72, TO-3, TO-18, TO-39, TO-92, TO-126, TO220, TO247, TO251, TO262, ZTX851

Flexible Transistors

Researchers have made several kinds of flexible transistors, including organic field-effect transistors. Flexible transistors are useful in some kinds of flexible displays and other flexible electronics.

Capacitor

A capacitor (originally known as a condenser) is a passive two-terminal electrical component used to temporarily store electrical energy in an electric field. The forms of practical capacitors vary widely, but most contain at least two electrical conductors (plates) separated by a dielectric. The conductors can be thin films, foils or sintered beads of metal or conductive electrolyte, etc. The nonconducting dielectric acts to increase the capacitor's charge capacity. Materials commonly used as dielectrics include glass, ceramic, plastic film, paper, mica, and oxide layers. Capacitors are widely used as parts of electrical circuits in many common electrical devices. Unlike a resistor, an ideal capacitor does not dissipate energy. Instead, a capacitor stores energy in the form of an electrostatic field between its plates.

When there is a potential difference across the conductors (e.g., when a capacitor is attached across a battery), an electric field develops across the dielectric, causing positive charge $+Q$ to collect on one plate and negative charge $-Q$ to collect on the other plate. If a battery has been attached to a capacitor for a sufficient amount of time, no current can flow through the capacitor. However, if a time-varying voltage is applied across the leads of the capacitor, a displacement current can flow.

An ideal capacitor is characterized by a single constant value, its capacitance. Capacitance is defined as the ratio of the electric charge Q on each conductor to the potential difference V between them. The SI unit of capacitance is the farad (F), which is equal to one coulomb per volt (1 C/V). Typical capacitance values range from about 1 pF (10^{-12} F) to about 1 mF (10^{-3} F).

The larger the surface area of the "plates" (conductors) and the narrower the gap between them, the greater the capacitance is. In practice, the dielectric between the plates passes a small amount of leakage current and also has an electric field strength limit, known as the breakdown voltage. The conductors and leads introduce an undesired inductance and resistance.

Capacitors are widely used in electronic circuits for blocking direct current while allowing alternating current to pass. In analog filter networks, they smooth the output of power supplies. In resonant circuits they tune radios to particular frequencies. In electric power transmission systems, they stabilize voltage and power flow.

History

Battery of four Leyden jars in Museum Boerhaave, Leiden, the Netherlands

In October 1745, Ewald Georg von Kleist of Pomerania, Germany, found that charge could be stored by connecting a high-voltage electrostatic generator by a wire to a volume of water in a hand-held glass jar. Von Kleist's hand and the water acted as conductors, and the jar as a dielectric (although details of the mechanism were incorrectly identified at the time). Von Kleist found that touching the wire resulted in a powerful spark, much more painful than that obtained from an electrostatic machine. The following year, the Dutch physicist Pieter van Musschenbroek invented a similar capacitor, which was named the Leyden jar, after the University of Leiden where he worked. He also was impressed by the power of the shock he received, writing, "I would not take a second shock for the kingdom of France."

Daniel Gralath was the first to combine several jars in parallel into a "battery" to increase the charge storage capacity. Benjamin Franklin investigated the Leyden jar and came to the conclusion that the charge was stored on the glass, not in the water as others had assumed. He also adopted the term "battery", (denoting the increasing of power with a row of similar units as in a battery of cannon), subsequently applied to clusters of electrochemical cells. Leyden jars were later made by coating the inside and outside of jars with metal foil, leaving a space at the mouth to prevent arcing between the foils. The earliest unit of capacitance was the jar, equivalent to about 1.11 nanofarads.

Leyden jars or more powerful devices employing flat glass plates alternating with foil conductors were used exclusively up until about 1900, when the invention of wireless (radio) created a demand for standard capacitors, and the steady move to higher frequencies required capacitors with lower inductance. More compact construction methods began to be used, such as a flexible dielectric sheet (like oiled paper) sandwiched between sheets of metal foil, rolled or folded into a small package.

Early capacitors were also known as *condensers*, a term that is still occasionally used today, particularly in high power applications, like automotive systems. The term was first used for this purpose by Alessandro Volta in 1782, with reference to the device's ability to store a higher density of electric charge than a normal isolated conductor.

Since the beginning of the study of electricity non conductive materials like glass, porcelain, paper and mica have been used as insulators. These materials some decades later were also well-suited for further use as the dielectric for the first capacitors. Paper capacitors made by sandwiching a strip of impregnated paper between strips of metal, and rolling the result into a cylinder were commonly used in the late 19century; their manufacture started in 1876, and they were used from the early 20th century as decoupling capacitors in telecommunications (telephony).

Porcelain was used in the first ceramic capacitors. In the early years of Marconi`s wireless transmitting apparatus porcelain capacitors were used for high voltage and high frequency application in the transmitters. On the receiver side smaller mica capacitors were used for resonant circuits. Mica dielectric capacitors were invented in 1909 by William Dubilier. Prior to World War II, mica was the most common dielectric for capacitors in the United States.

Charles Pollak (born Karol Pollak), the inventor of the first electrolytic capacitors, found out that the oxide layer on an aluminum anode remained stable in a neutral or alkaline electrolyte, even when the power was switched off. In 1896 he was granted U.S. Patent No. 672,913 for an "Electric liquid capacitor with aluminum electrodes." Solid electrolyte tantalum capacitors were invented by Bell Laboratories in the early 1950s as a miniaturized and more reliable low-voltage support capacitor to complement their newly invented transistor.

With the development of plastic materials by organic chemists during the Second World War, the capacitor industry began to replace paper with thinner polymer films. One very early development in film capacitors was described in British Patent 587,953 in 1944.

Last but not least the electric double-layer capacitor (now Supercapacitors) were invented. In 1957 H. Becker developed a "Low voltage electrolytic capacitor with porous carbon electrodes". He believed that the energy was stored as a charge in the carbon pores used in his capacitor as in the pores of the etched foils of electrolytic capacitors. Because the double layer mechanism was not known by him at the time, he wrote in the patent: "It is not known exactly what is taking place in the component if it is used for energy storage, but it leads to an extremely high capacity.".

Theory of Operation

Overview

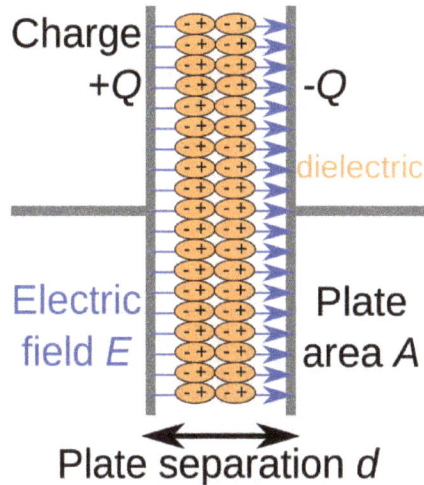

Charge separation in a parallel-plate capacitor causes an internal electric field.
A dielectric (orange) reduces the field and increases the capacitance.

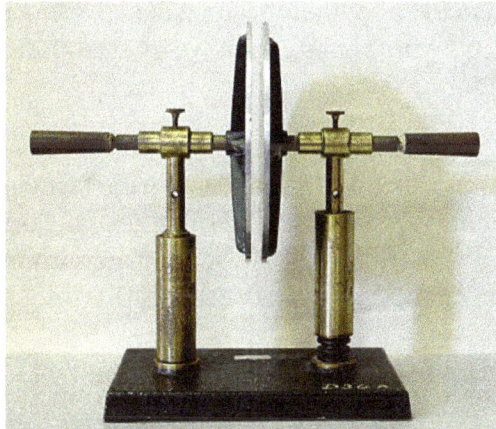

A simple demonstration capacitor made of two parallel metal plates, using an air gap as the dielectric.

A capacitor consists of two conductors separated by a non-conductive region. The non-conductive region can either be a vacuum or an electrical insulator material known as a dielectric. Examples of dielectric media are glass, air, paper, and even a semiconductor depletion region chemically identical to the conductors. A capacitor is assumed to be self-contained and isolated, with no net electric charge and no influence from any external electric field. The conductors thus hold equal and opposite charges on their facing surfaces, and the dielectric develops an electric field. In SI units, a capacitance of one farad means that one coulomb of charge on each conductor causes a voltage of one volt across the device.

An ideal capacitor is wholly characterized by a constant capacitance C, defined as the ratio of charge $\pm Q$ on each conductor to the voltage V between them:

$$C = \frac{Q}{V}$$

Because the conductors (or plates) are close together, the opposite charges on the conductors attract one another due to their electric fields, allowing the capacitor to store more charge for a given voltage than if the conductors were separated, giving the capacitor a large capacitance.

Sometimes charge build-up affects the capacitor mechanically, causing its capacitance to vary. In this case, capacitance is defined in terms of incremental changes:

$$C = \frac{dQ}{dV}$$

Hydraulic Analogy

In the hydraulic analogy, a capacitor is analogous to a rubber membrane sealed inside a pipe. This animation illustrates a membrane being repeatedly stretched and un-stretched by the flow of water, which is analogous to a capacitor being repeatedly charged and discharged by the flow of charge.

In the hydraulic analogy, charge carriers flowing through a wire are analogous to water flowing through a pipe. A capacitor is like a rubber membrane sealed inside a pipe. Water molecules cannot pass through the membrane, but some water can move by stretching the membrane. The analogy clarifies a few aspects of capacitors:

- *The current alters the charge on a capacitor*, just as the flow of water changes the position of the membrane. More specifically, the effect of an electric current is to increase the charge of one plate of the capacitor, and decrease the charge of the other plate by an equal amount. This is just as when water flow moves the rubber membrane, it increases the amount of water on one side of the membrane, and decreases the amount of water on the other side.

- *The more a capacitor is charged, the larger its voltage drop*; i.e., the more it "pushes back" against the charging current. This is analogous to the fact that the more a membrane is stretched, the more it pushes back on the water.

- *Charge can flow "through" a capacitor even though no individual electron can get from one side to the other*. This is analogous to the fact that water can flow through the pipe even though no water molecule can pass through the rubber membrane. Of course, the flow cannot continue in the same direction forever; the capacitor will experience dielectric breakdown, and analogously the membrane will eventually break.

- The *capacitance* describes how much charge can be stored on one plate of a capacitor for a given "push" (voltage drop). A very stretchy, flexible membrane corresponds to a higher capacitance than a stiff membrane.

- A charged-up capacitor is storing potential energy, analogously to a stretched membrane.

Energy of Electric Field

Work must be done by an external influence to "move" charge between the conductors in a capacitor. When the external influence is removed, the charge separation persists in the electric field and energy is stored to be released when the charge is allowed to return to its equilibrium position. The work done in establishing the electric field, and hence the amount of energy stored, is

$$W = \int_0^Q V(q)\mathrm{d}q = \int_0^Q \frac{q}{C}\mathrm{d}q = \frac{1}{2}\frac{Q^2}{C} = \frac{1}{2}CV^2 = \frac{1}{2}VQ$$

Here Q is the charge stored in the capacitor, V is the voltage across the capacitor, and C is the capacitance.

In the case of a fluctuating voltage $V(t)$, the stored energy also fluctuates and hence power must flow into or out of the capacitor. This power can be found by taking the time derivative of the stored energy:

$$P = \frac{\mathrm{d}W}{\mathrm{d}t} = \frac{\mathrm{d}}{\mathrm{d}t}\left(\frac{1}{2}CV^2\right) = CV(t)\frac{\mathrm{d}V(t)}{\mathrm{d}t}$$

A real capacitor with loss may be modeled as an ideal capacitor that has an Equivalent Series Resistance (ESR) which dissipates power as the capacitor is charged or discharged. For an sinusoidal input voltage the power dissipated due to the ESR is given as:

$$Pd_{rms} = \frac{V_{rms}^2 R_{esr}}{R_{esr}^2 + \left(\dfrac{1}{2\pi fC}\right)^2}$$

Current–voltage Relation

The current $I(t)$ through any component in an electric circuit is defined as the rate of flow of a charge $Q(t)$ passing through it, but actual charges—electrons—cannot pass through the dielectric layer of a capacitor. Rather, one electron accumulates on the negative plate for each one that leaves the positive plate, resulting in an electron depletion and consequent positive charge on one electrode that is equal and opposite to the accumulated negative charge on the other. Thus the charge on the electrodes is equal to the integral of the current as well as proportional to the voltage, as discussed above. As with any antiderivative, a constant of integration is added to represent the initial voltage $V(t_0)$. This is the integral form of the capacitor equation:

$$V(t) = \frac{Q(t)}{C} = \frac{1}{C}\int_{t_0}^t I(\tau)\mathrm{d}\tau + V(t_0)$$

Taking the derivative of this and multiplying by C yields the derivative form:

$$I(t) = \frac{\mathrm{d}Q(t)}{\mathrm{d}t} = C\frac{\mathrm{d}V(t)}{\mathrm{d}t}$$

The dual of the capacitor is the inductor, which stores energy in a magnetic field rather than an electric field. Its current-voltage relation is obtained by exchanging current and voltage in the capacitor equations and replacing C with the inductance L.

DC Circuits

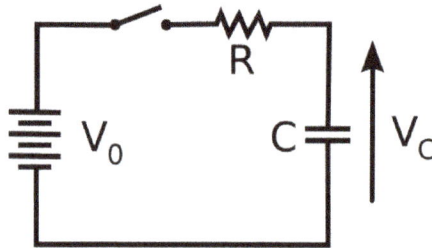

A simple resistor-capacitor circuit demonstrates charging of a capacitor.

A series circuit containing only a resistor, a capacitor, a switch and a constant DC source of voltage V_0 is known as a *charging circuit*. If the capacitor is initially uncharged while the switch is open, and the switch is closed at t_0, it follows from Kirchhoff's voltage law that

$$V_0 = v_{resistor}(t) + v_{capacitor}(t) = i(t)R + \frac{1}{C}\int_{t_0}^{t} i(\tau)\mathrm{d}\tau$$

Taking the derivative and multiplying by C, gives a first-order differential equation:

$$RC\frac{di(t)}{dt} + i(t) = 0$$

At $t = 0$, the voltage across the capacitor is zero and the voltage across the resistor is V_0. The initial current is then $I(0) = V_0/R$. With this assumption, solving the differential equation yields

$$I(t) = \frac{V_0}{R}e^{-\frac{t}{\tau_0}}$$

$$V(t) = V_0\left(1 - e^{-\frac{t}{\tau_0}}\right)$$

where $\tau_0 = RC$ is the *time constant* of the system. As the capacitor reaches equilibrium with the source voltage, the voltages across the resistor and the current through the entire circuit decay exponentially. The case of *discharging* a charged capacitor likewise demonstrates exponential decay, but with the initial capacitor voltage replacing V_0 and the final voltage being zero.

AC Circuits

Impedance, the vector sum of reactance and resistance, describes the phase difference and the ratio of amplitudes between sinusoidally varying voltage and sinusoidally varying current at a given frequency. Fourier analysis allows any signal to be constructed from a spectrum of frequencies, whence the circuit's reaction to the various frequencies may be found. The reactance and impedance of a capacitor are respectively

$$X = -\frac{1}{\omega C} = -\frac{1}{2\pi fC}$$

$$Z = \frac{1}{j\omega C} = -\frac{j}{\omega C} = -\frac{j}{2\pi fC}$$

where j is the imaginary unit and ω is the angular frequency of the sinusoidal signal. The $-j$ phase indicates that the AC voltage $V = ZI$ lags the AC current by 90°: the positive current phase corresponds to increasing voltage as the capacitor charges; zero current corresponds to instantaneous constant voltage, etc.

Impedance decreases with increasing capacitance and increasing frequency. This implies that a higher-frequency signal or a larger capacitor results in a lower voltage amplitude per current amplitude—an AC "short circuit" or AC coupling. Conversely, for very low frequencies, the reactance will be high, so that a capacitor is nearly an open circuit in AC analysis—those frequencies have been "filtered out".

Capacitors are different from resistors and inductors in that the impedance is *inversely* proportional to the defining characteristic; i.e., capacitance.

A capacitor connected to a sinusoidal voltage source will cause a displacement current to flow through it. In the case that the voltage source is $V_0\cos(\omega t)$, the displacement current can be expressed as:

$$I = C\frac{dV}{dt} = -\omega CV_0 \sin(\omega t)$$

At $\sin(\omega t) = -1$, the capacitor has a maximum (or peak) current whereby $I_0 = \omega CV_0$. The ratio of peak voltage to peak current is due to capacitive reactance (denoted X_C).

$$X_C = \frac{V_0}{I_0} = \frac{V_0}{\omega CV_0} = \frac{1}{\omega C}$$

X_C approaches zero as ω approaches infinity. If X_C approaches 0, the capacitor resembles a short wire that strongly passes current at high frequencies. X_C approaches infinity as ω approaches zero. If X_C approaches infinity, the capacitor resembles an open circuit that poorly passes low frequencies.

The current of the capacitor may be expressed in the form of cosines to better compare with the voltage of the source:

$$I = -I_0 \sin(\omega t) = I_0 \cos(\omega t + 90°)$$

In this situation, the current is out of phase with the voltage by $+\pi/2$ radians or +90 degrees (i.e., the current will lead the voltage by 90°).

Laplace Circuit Analysis (s-domain)

When using the Laplace transform in circuit analysis, the impedance of an ideal capacitor with no initial charge is represented in the s domain by:

$$Z(s) = \frac{1}{sC}$$

where

- C is the capacitance, and

- s is the complex frequency.

Parallel-plate Model

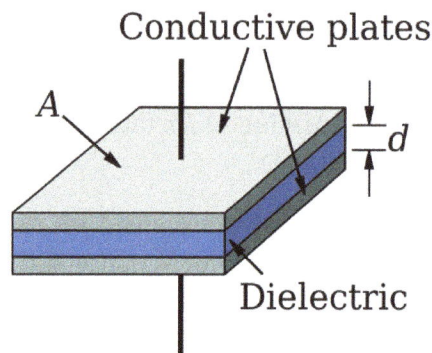

Conductive plates

A

d

Dielectric

Dielectric is placed between two conducting plates, each of area A and with a separation of d

The simplest model capacitor consists of two thin parallel conductive plates separated by a dielectric with permittivity ε. This model may also be used to make qualitative predictions for other device geometries. The plates are considered to extend uniformly over an area A and a charge density $\pm\rho$ = $\pm Q/A$ exists on their surface. Assuming that the length and width of the plates are much greater than their separation d, the electric field near the centre of the device will be uniform with the magnitude $E = \rho/\varepsilon$. The voltage is defined as the line integral of the electric field between the plates

$$V = \int_0^d E\,\mathrm{d}z = \int_0^d \frac{\rho}{\varepsilon}\,\mathrm{d}z = \frac{\rho d}{\varepsilon} = \frac{Qd}{\varepsilon A}$$

Solving this for $C = Q/V$ reveals that capacitance increases with area of the plates, and decreases as separation between plates increases.

$$C = \frac{\varepsilon A}{d}$$

The capacitance is therefore greatest in devices made from materials with a high permittivity, large plate area, and small distance between plates.

A parallel plate capacitor can only store a finite amount of energy before dielectric breakdown occurs. The capacitor's dielectric material has a dielectric strength U_d which sets the capacitor's breakdown voltage at $V = V_{bd} = U_d d$. The maximum energy that the capacitor can store is therefore

$$E = \frac{1}{2}CV^2 = \frac{1}{2}\frac{\varepsilon A}{d}(U_d d)^2 = \frac{1}{2}\varepsilon A d U_d^2$$

The maximum energy is a function of dielectric volume, permittivity, and dielectric strength. Changing the plate area and the separation between the plates while maintaining the same volume causes no change of the maximum amount of energy that the capacitor can store, so long as the distance between plates remains much smaller than both the length and width of the plates. In addition, these equations assume that the electric field is entirely concentrated in the dielectric between the plates. In reality there are fringing fields outside the dielectric, for example between the sides of the capacitor plates, which will increase the effective capacitance of the capacitor. This is sometimes called parasitic capacitance. For some simple capacitor geometries this additional capacitance term can be calculated analytically. It becomes negligibly small when the ratios of plate width to separation and length to separation are large.

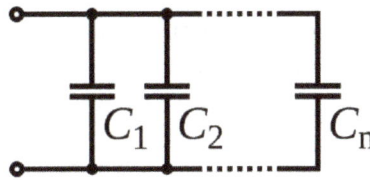

Several capacitors in parallel

Networks

For capacitors in parallel

> Capacitors in a parallel configuration each have the same applied voltage. Their capacitances add up. Charge is apportioned among them by size. Using the schematic diagram to visualize parallel plates, it is apparent that each capacitor contributes to the total surface area.

$$C_{eq} = C_1 + C_2 + \cdots + C_n$$

For capacitors in series

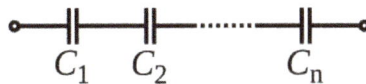

Several capacitors in series

> Connected in series, the schematic diagram reveals that the separation distance, not the plate area, adds up. The capacitors each store instantaneous charge build-up equal to that of every other capacitor in the series. The total voltage difference from end to end is apportioned to each capacitor according to the inverse of its capacitance. The entire series acts as a capacitor *smaller* than any of its components.

$$\frac{1}{C_{eq}} = \frac{1}{C_1} + \frac{1}{C_2} + \cdots + \frac{1}{C_n}$$

Capacitors are combined in series to achieve a higher working voltage, for example for smoothing a high voltage power supply. The voltage ratings, which are based on plate separation, add up, if capacitance and leakage currents for each capacitor are identical. In such an application, on occasion, series strings are connected in parallel, forming a matrix. The goal is to maximize the energy storage of the network without overloading any capacitor. For high-energy storage with capacitors in series, some safety considerations must be applied to ensure one capacitor failing and leaking current will not apply too much voltage to the other series capacitors.

Series connection is also sometimes used to adapt polarized electrolytic capacitors for bipolar AC use.

Voltage distribution in parallel-to-series networks.

To model the distribution of voltages from a single charged capacitor (A) connected in parallel to a chain of capacitors in series (B_n) :

$$(volts)A_{eq} = A\left(1 - \frac{1}{n+1}\right)$$

$$(volts)B_{1..n} = \frac{A}{n}\left(1 - \frac{1}{n+1}\right)$$

$$A - B = 0$$

Note: This is only correct if all capacitance values are equal.

The power transferred in this arrangement is:

$$P = \frac{1}{R} \cdot \frac{1}{n+1} A_{volts}\left(A_{farads} + B_{farads}\right)$$

Non-ideal Behavior

Capacitors deviate from the ideal capacitor equation in a number of ways. Some of these, such as leakage current and parasitic effects are linear, or can be analyzed as nearly linear, and can be dealt with by adding virtual components to the equivalent circuit of an ideal capacitor. The usual methods of network analysis can then be applied. In other cases, such as with breakdown voltage, the effect is non-linear and ordinary (normal, eg, linear) network analysis cannot be used, the effect must be dealt with separately. There is yet another group, which may be linear but invalidate the assumption in the analysis that capacitance is a constant. Such an example is temperature dependence. Finally, combined parasitic effects such as inherent inductance, resistance, or dielectric losses can exhibit non-uniform behavior at variable frequencies of operation.

Breakdown Voltage

Above a particular electric field, known as the dielectric strength E_{ds}, the dielectric in a capacitor becomes conductive. The voltage at which this occurs is called the breakdown voltage of the device, and is given by the product of the dielectric strength and the separation between the conductors,

$$V_{bd} = E_{ds}d$$

The maximum energy that can be stored safely in a capacitor is limited by the breakdown voltage. Due to the scaling of capacitance and breakdown voltage with dielectric thickness, all capacitors made with a particular dielectric have approximately equal maximum energy density, to the extent that the dielectric dominates their volume.

For air dielectric capacitors the breakdown field strength is of the order 2 to 5 MV/m; for mica the breakdown is 100 to 300 MV/m; for oil, 15 to 25 MV/m; it can be much less when other materials are used for the dielectric. The dielectric is used in very thin layers and so absolute breakdown voltage of capacitors is limited. Typical ratings for capacitors used for general electronics applications range from a few volts to 1 kV. As the voltage increases, the dielectric must be thicker, making high-voltage capacitors larger per capacitance than those rated for lower voltages. The breakdown voltage is critically affected by factors such as the geometry of the capacitor conductive parts; sharp edges or points increase the electric field strength at that point and can lead to a local breakdown. Once this starts to happen, the breakdown quickly tracks through the dielectric until it reaches the opposite plate, leaving carbon behind and causing a short (or relatively low resistance) circuit. The results can be explosive as the short in the capacitor draws current from the surrounding circuitry and dissipates the energy.

The usual breakdown route is that the field strength becomes large enough to pull electrons in the dielectric from their atoms thus causing conduction. Other scenarios are possible, such as impurities in the dielectric, and, if the dielectric is of a crystalline nature, imperfections in the crystal structure can result in an avalanche breakdown as seen in semi-conductor devices. Breakdown voltage is also affected by pressure, humidity and temperature.

Equivalent Circuit

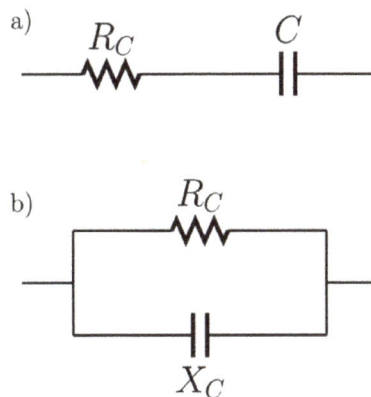

Two different circuit models of a real capacitor

An ideal capacitor only stores and releases electrical energy, without dissipating any. In reality, all capacitors have imperfections within the capacitor's material that create resistance. This is specified as the *equivalent series resistance* or **ESR** of a component. This adds a real component to the impedance:

$$Z_C = Z + R_{ESR} = \frac{1}{j\omega C} + R_{ESR}$$

As frequency approaches infinity, the capacitive impedance (or reactance) approaches zero and the ESR becomes significant. As the reactance becomes negligible, power dissipation approaches $P_{RMS} = V_{RMS}^2 / R_{ESR}$.

Similarly to ESR, the capacitor's leads add *equivalent series inductance* or **ESL** to the component. This is usually significant only at relatively high frequencies. As inductive reactance is positive and increases with frequency, above a certain frequency capacitance will be canceled by inductance. High-frequency engineering involves accounting for the inductance of all connections and components.

If the conductors are separated by a material with a small conductivity rather than a perfect dielectric, then a small leakage current flows directly between them. The capacitor therefore has a finite parallel resistance, and slowly discharges over time (time may vary greatly depending on the capacitor material and quality).

Q factor

The quality factor (or Q) of a capacitor is the ratio of its reactance to its resistance at a given frequency, and is a measure of its efficiency. The higher the Q factor of the capacitor, the closer it approaches the behavior of an ideal, lossless, capacitor.

The Q factor of a capacitor can be found through the following formula:

$$Q = \frac{X_C}{R_C} = \frac{1}{\omega C R_C},$$

where ω is angular frequency, C is the capacitance, X_C is the capacitive reactance, and R_C is the series resistance of the capacitor.

Ripple Current

Ripple current is the AC component of an applied source (often a switched-mode power supply) whose frequency may be constant or varying. Ripple current causes heat to be generated within the capacitor due to the dielectric losses caused by the changing field strength together with the current flow across the slightly resistive supply lines or the electrolyte in the capacitor. The equivalent series resistance (ESR) is the amount of internal series resistance one would add to a perfect capacitor to model this.

Some types of capacitors, primarily tantalum and aluminum electrolytic capacitors, as well as some film capacitors have a specified rating value for maximum ripple current.

- Tantalum electrolytic capacitors with solid manganese dioxide electrolyte are limited by ripple current and generally have the highest ESR ratings in the capacitor family. Exceeding their ripple limits can lead to shorts and burning parts.

- Aluminum electrolytic capacitors, the most common type of electrolytic, suffer a shortening of life expectancy at higher ripple currents. If ripple current exceeds the rated value of the capacitor, it tends to result in explosive failure.

- Ceramic capacitors generally have no ripple current limitation and have some of the lowest ESR ratings.

- Film capacitors have very low ESR ratings but exceeding rated ripple current may cause degradation failures.

Capacitance Instability

The capacitance of certain capacitors decreases as the component ages. In ceramic capacitors, this is caused by degradation of the dielectric. The type of dielectric, ambient operating and storage temperatures are the most significant aging factors, while the operating voltage has a smaller effect. The aging process may be reversed by heating the component above the Curie point. Aging is fastest near the beginning of life of the component, and the device stabilizes over time. Electrolytic capacitors age as the electrolyte evaporates. In contrast with ceramic capacitors, this occurs towards the end of life of the component.

Temperature dependence of capacitance is usually expressed in parts per million (ppm) per °C. It can usually be taken as a broadly linear function but can be noticeably non-linear at the temperature extremes. The temperature coefficient can be either positive or negative, sometimes even amongst different samples of the same type. In other words, the spread in the range of temperature coefficients can encompass zero.

Capacitors, especially ceramic capacitors, and older designs such as paper capacitors, can absorb sound waves resulting in a microphonic effect. Vibration moves the plates, causing the capacitance to vary, in turn inducing AC current. Some dielectrics also generate piezoelectricity. The resulting interference is especially problematic in audio applications, potentially causing feedback or unintended recording. In the reverse microphonic effect, the varying electric field between the capacitor plates exerts a physical force, moving them as a speaker. This can generate audible sound, but drains energy and stresses the dielectric and the electrolyte, if any.

Current and Voltage Reversal

Current reversal occurs when the current changes direction. Voltage reversal is the change of polarity in a circuit. Reversal is generally described as the percentage of the maximum rated voltage that reverses polarity. In DC circuits, this will usually be less than 100% (often in the range of 0 to 90%), whereas AC circuits experience 100% reversal.

In DC circuits and pulsed circuits, current and voltage reversal are affected by the damping of the system. Voltage reversal is encountered in RLC circuits that are under-damped. The current and voltage reverse direction, forming a harmonic oscillator between the inductance and capacitance.

The current and voltage will tend to oscillate and may reverse direction several times, with each peak being lower than the previous, until the system reaches an equilibrium. This is often referred to as ringing. In comparison, critically damped or over-damped systems usually do not experience a voltage reversal. Reversal is also encountered in AC circuits, where the peak current will be equal in each direction.

For maximum life, capacitors usually need to be able to handle the maximum amount of reversal that a system will experience. An AC circuit will experience 100% voltage reversal, while under-damped DC circuits will experience less than 100%. Reversal creates excess electric fields in the dielectric, causes excess heating of both the dielectric and the conductors, and can dramatically shorten the life expectancy of the capacitor. Reversal ratings will often affect the design considerations for the capacitor, from the choice of dielectric materials and voltage ratings to the types of internal connections used.

Dielectric Absorption

Capacitors made with any type of dielectric material will show some level of "dielectric absorption" or "soakage". On discharging a capacitor and disconnecting it, after a short time it may develop a voltage due to hysteresis in the dielectric. This effect is objectionable in applications such as precision sample and hold circuits or timing circuits. The level of absorption depends on many factors, from design considerations to charging time, since the absorption is a time-dependent process. However, the primary factor is the type of dielectric material. Capacitors such as tantalum electrolytic or polysulfone film exhibit relatively high absorption, while polystyrene or Teflon allow very small levels of absorption. In some capacitors where dangerous voltages and energies exist, such as in flashtubes, television sets, and defibrillators, the dielectric absorption can recharge the capacitor to hazardous voltages after it has been shorted or discharged. Any capacitor containing over 10 joules of energy is generally considered hazardous, while 50 joules or higher is potentially lethal. A capacitor may regain anywhere from 0.01 to 20% of its original charge over a period of several minutes, allowing a seemingly safe capacitor to become surprisingly dangerous.

Leakage

Leakage is equivalent to a resistor in parallel with the capacitor. Constant exposure to heat can cause dielectric breakdown and excessive leakage, a problem often seen in older vacuum tube circuits, particularly where oiled paper and foil capacitors were used. In many vacuum tube circuits, interstage coupling capacitors are used to conduct a varying signal from the plate of one tube to the grid circuit of the next stage. A leaky capacitor can cause the grid circuit voltage to be raised from its normal bias setting, causing excessive current or signal distortion in the downstream tube. In power amplifiers this can cause the plates to glow red, or current limiting resistors to overheat, even fail. Similar considerations apply to component fabricated solid-state (transistor) amplifiers, but owing to lower heat production and the use of modern polyester dielectric barriers this once-common problem has become relatively rare.

Electrolytic Failure from Disuse

Aluminum electrolytic capacitors are *conditioned* when manufactured by applying a voltage sufficient to initiate the proper internal chemical state. This state is maintained by regular use of the equipment. If a system using electrolytic capacitors is unused for a long period of time it can lose its conditioning. Sometimes they fail with a short circuit when next operated.

Capacitor Types

Practical capacitors are available commercially in many different forms. The type of internal dielectric, the structure of the plates and the device packaging all strongly affect the characteristics of the capacitor, and its applications.

Values available range from very low (picofarad range; while arbitrarily low values are in principle possible, stray (parasitic) capacitance in any circuit is the limiting factor) to about 5 kF supercapacitors.

Above approximately 1 microfarad electrolytic capacitors are usually used because of their small size and low cost compared with other types, unless their relatively poor stability, life and polarised nature make them unsuitable. Very high capacity supercapacitors use a porous carbon-based electrode material.

Dielectric Materials

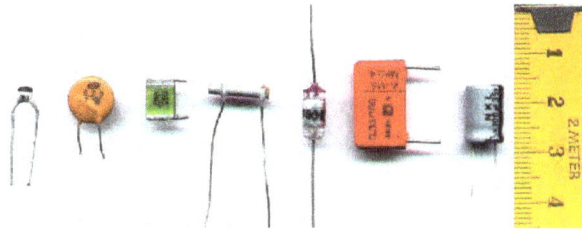

Capacitor materials. From left: multilayer ceramic, ceramic disc, multilayer polyester film, tubular ceramic, polystyrene, metalized polyester film, aluminum electrolytic. Major scale divisions are in centimetres.

Most capacitors have a dielectric spacer, which increases their capacitance compared to air or a vacuum. In order to maximise the charge that a capacitor can hold, the dielectric material needs to have as high a permittivity as possible, while also having as high a breakdown voltage as possible. The dielectric also need to have as lower loss with frequency as possible

However, low value capacitors are available with a vacuum between their plates to allow extremely high voltage operation and low losses. Variable capacitors with their plates open to the atmosphere were commonly used in radio tuning circuits. Later designs use polymer foil dielectric between the moving and stationary plates, with no significant air space between the plates.

Several solid dielectrics are available, including paper, plastic, glass, mica and ceramic.

Paper was used extensively in older capacitors and offers relatively high voltage performance. However, paper absorbs moisture, and has been largely replaced by plastic film capacitors.

Most of the plastic films now used offer better stability and ageing performance than such older dielectrics such as oiled paper, which makes them useful in timer circuits, although they may be limited to relatively low operating temperatures and frequencies, because of the limitations of the plastic film being used. Large plastic film capacitors are used extensively in suppression circuits, motor start circuits, and power factor correction circuits.

Ceramic capacitors are generally small, cheap and useful for high frequency applications, although their capacitance varies strongly with voltage and temperature and they age poorly. They can also suffer from the piezoelectric effect. Ceramic capacitors are broadly categorized as class 1 dielectrics, which have predictable variation of capacitance with temperature or class 2 dielectrics, which can operate at higher voltage. Modern multilayer ceramics are usually quite small, but some types have inherently wide value tolerances, microphonic issues, and are usually physically brittle.

Glass and mica capacitors are extremely reliable, stable and tolerant to high temperatures and voltages, but are too expensive for most mainstream applications.

Electrolytic capacitors and supercapacitors are used to store small and larger amounts of energy, respectively, ceramic capacitors are often used in resonators, and parasitic capacitance occurs in circuits wherever the simple conductor-insulator-conductor structure is formed unintentionally by the configuration of the circuit layout.

Electrolytic capacitors use an aluminum or tantalum plate with an oxide dielectric layer. The second electrode is a liquid electrolyte, connected to the circuit by another foil plate. Electrolytic capacitors offer very high capacitance but suffer from poor tolerances, high instability, gradual loss of capacitance especially when subjected to heat, and high leakage current. Poor quality capacitors may leak electrolyte, which is harmful to printed circuit boards. The conductivity of the electrolyte drops at low temperatures, which increases equivalent series resistance. While widely used for power-supply conditioning, poor high-frequency characteristics make them unsuitable for many applications. Electrolytic capacitors will self-degrade if unused for a period (around a year), and when full power is applied may short circuit, permanently damaging the capacitor and usually blowing a fuse or causing failure of rectifier diodes (for instance, in older equipment, arcing in rectifier tubes). They can be restored before use (and damage) by gradually applying the operating voltage, often done on antique vacuum tube equipment over a period of 30 minutes by using a variable transformer to supply AC power. Unfortunately, the use of this technique may be less satisfactory for some solid state equipment, which may be damaged by operation below its normal power range, requiring that the power supply first be isolated from the consuming circuits. Such remedies may not be applicable to modern high-frequency power supplies as these produce full output voltage even with reduced input.

Tantalum capacitors offer better frequency and temperature characteristics than aluminum, but higher dielectric absorption and leakage.

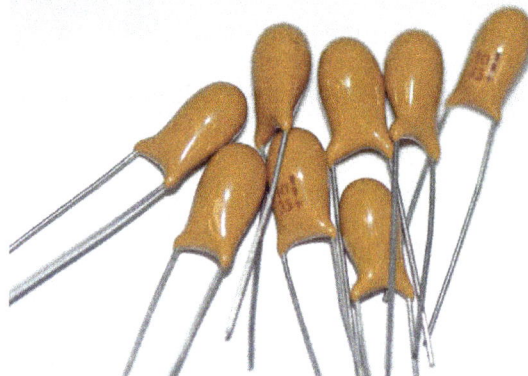

Solid electrolyte, resin-dipped 10 µF 35 V tantalum capacitors. The + sign indicates the positive lead.

Polymer capacitors (OS-CON, OC-CON, KO, AO) use solid conductive polymer (or polymerized organic semiconductor) as electrolyte and offer longer life and lower ESR at higher cost than standard electrolytic capacitors.

A feedthrough capacitor is a component that, while not serving as its main use, has capacitance and is used to conduct signals through a conductive sheet.

Several other types of capacitor are available for specialist applications. Supercapacitors store large amounts of energy. Supercapacitors made from carbon aerogel, carbon nanotubes, or highly porous electrode materials, offer extremely high capacitance (up to 5 kF as of 2010) and can be used in some applications instead of rechargeable batteries. Alternating current capacitors are specifically designed to work on line (mains) voltage AC power circuits. They are commonly used in electric motor circuits and are often designed to handle large currents, so they tend to be physically large. They are usually ruggedly packaged, often in metal cases that can be easily grounded/earthed. They also are designed with direct current breakdown voltages of at least five times the maximum AC voltage.

Voltage-dependent Capacitors

The dielectric constant for a number of very useful dielectrics changes as a function of the applied electrical field, for example ferroelectric materials, so the capacitance for these devices is more complex. For example, in charging such a capacitor the differential increase in voltage with charge is governed by:

$$dQ = C(V)dV$$

where the voltage dependence of capacitance, $C(V)$, suggests that the capacitance is a function of the electric field strength, which in a large area parallel plate device is given by $\varepsilon = V/d$. This field polarizes the dielectric, which polarization, in the case of a ferroelectric, is a nonlinear S-shaped function of the electric field, which, in the case of a large area parallel plate device, translates into a capacitance that is a nonlinear function of the voltage.

Corresponding to the voltage-dependent capacitance, to charge the capacitor to voltage V an integral relation is found:

$$Q = \int_0^V C(V)dV$$

which agrees with $Q = CV$ only when C does not depend on voltage V.

By the same token, the energy stored in the capacitor now is given by

$$dW = QdV = \left[\int_0^V dV'\, C(V') \right] dV .$$

Integrating:

$$W = \int_0^V dV \int_0^V dV'\, C(V') = \int_0^V dV' \int_{V'}^V dV\, C(V') = \int_0^V dV'(V - V')C(V'),$$

where interchange of the order of integration is used.

The nonlinear capacitance of a microscope probe scanned along a ferroelectric surface is used to study the domain structure of ferroelectric materials.

Another example of voltage dependent capacitance occurs in semiconductor devices such as semiconductor diodes, where the voltage dependence stems not from a change in dielectric constant but in a voltage dependence of the spacing between the charges on the two sides of the capacitor. This effect is intentionally exploited in diode-like devices known as varicaps.

Frequency-dependent Capacitors

If a capacitor is driven with a time-varying voltage that changes rapidly enough, at some frequency the polarization of the dielectric cannot follow the voltage. As an example of the origin of this mechanism, the internal microscopic dipoles contributing to the dielectric constant cannot move instantly, and so as frequency of an applied alternating voltage increases, the dipole response is limited and the dielectric constant diminishes. A changing dielectric constant with frequency is referred to as dielectric dispersion, and is governed by dielectric relaxation processes, such as Debye relaxation. Under transient conditions, the displacement field can be expressed as:

$$\mathbf{D}(\mathbf{t}) = \varepsilon_0 \int_{-\infty}^{t} \varepsilon_r(t-t')\mathbf{E}(t')\,dt',$$

indicating the lag in response by the time dependence of ε_r, calculated in principle from an underlying microscopic analysis, for example, of the dipole behavior in the dielectric. See, for example, linear response function. The integral extends over the entire past history up to the present time. A Fourier transform in time then results in:

$$\mathbf{D}(\omega) = \varepsilon_0 \varepsilon_r(\omega)\mathbf{E}(\omega),$$

where $\varepsilon_r(\omega)$ is now a complex function, with an imaginary part related to absorption of energy from the field by the medium. The capacitance, being proportional to the dielectric constant, also exhibits this frequency behavior. Fourier transforming Gauss's law with this form for displacement field:

$$I(\omega) = j\omega Q(\omega) = j\omega \oint_{\Sigma} D(r,\omega)\cdot d\Sigma$$

$$= \left[G(\omega) + j\omega C(\omega)\right]V(\omega) = \frac{V(\omega)}{Z(\omega)},$$

where j is the imaginary unit, $V(\omega)$ is the voltage component at angular frequency ω, $G(\omega)$ is the *real* part of the current, called the *conductance*, and $C(\omega)$ determines the *imaginary* part of the current and is the *capacitance*. $Z(\omega)$ is the complex impedance.

When a parallel-plate capacitor is filled with a dielectric, the measurement of dielectric properties of the medium is based upon the relation:

$$\varepsilon_r(\omega) = \varepsilon_r'(\omega) - j\varepsilon_r''(\omega) = \frac{1}{j\omega Z(\omega)C_0} = \frac{C_{cmplx}(\omega)}{C_0},$$

where a single *prime* denotes the real part and a double *prime* the imaginary part, $Z(\omega)$ is the complex impedance with the dielectric present, $C_{cmplx}(\omega)$ is the so-called *complex* capacitance with the dielectric present, and C_0 is the capacitance without the dielectric. (Measurement "without the dielectric" in principle means measurement in free space, an unattainable goal inasmuch as even the quantum vacuum is predicted to exhibit nonideal behavior, such as dichroism. For practical purposes, when measurement errors are taken into account, often a measurement in terrestrial vacuum, or simply a calculation of C_0, is sufficiently accurate.)

Using this measurement method, the dielectric constant may exhibit a resonance at certain frequencies corresponding to characteristic response frequencies (excitation energies) of contributors to the dielectric constant. These resonances are the basis for a number of experimental techniques for detecting defects. The *conductance method* measures absorption as a function of frequency. Alternatively, the time response of the capacitance can be used directly, as in *deep-level transient spectroscopy*.

Another example of frequency dependent capacitance occurs with MOS capacitors, where the slow generation of minority carriers means that at high frequencies the capacitance measures only the majority carrier response, while at low frequencies both types of carrier respond.

At optical frequencies, in semiconductors the dielectric constant exhibits structure related to the band structure of the solid. Sophisticated modulation spectroscopy measurement methods based upon modulating the crystal structure by pressure or by other stresses and observing the related changes in absorption or reflection of light have advanced our knowledge of these materials.

Structure

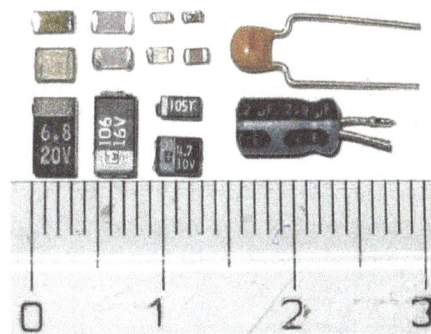

Capacitor packages: SMD ceramic at top left; SMD tantalum at bottom left; through-hole tantalum at top right; through-hole electrolytic at bottom right. Major scale divisions are cm.

The arrangement of plates and dielectric has many variations depending on the desired ratings of the capacitor. For small values of capacitance (microfarads and less), ceramic disks use metallic coatings, with wire leads bonded to the coating. Larger values can be made by multiple stacks of plates and disks. Larger value capacitors usually use a metal foil or metal film layer deposited on the surface of a dielectric film to make the plates, and a dielectric film of impregnated paper or plastic – these are rolled up to save space. To reduce the series resistance and inductance for long plates, the plates and dielectric are staggered so that connection is made at the common edge of the rolled-up plates, not at the ends of the foil or metalized film strips that comprise the plates.

The assembly is encased to prevent moisture entering the dielectric – early radio equipment used a cardboard tube sealed with wax. Modern paper or film dielectric capacitors are dipped in a hard thermoplastic. Large capacitors for high-voltage use may have the roll form compressed to fit into a rectangular metal case, with bolted terminals and bushings for connections. The dielectric in larger capacitors is often impregnated with a liquid to improve its properties.

Several axial-lead electrolytic capacitors

Capacitors may have their connecting leads arranged in many configurations, for example axially or radially. "Axial" means that the leads are on a common axis, typically the axis of the capacitor's cylindrical body – the leads extend from opposite ends. Radial leads might more accurately be referred to as tandem; they are rarely actually aligned along radii of the body's circle, so the term is inexact, although universal. The leads (until bent) are usually in planes parallel to that of the flat body of the capacitor, and extend in the same direction; they are often parallel as manufactured.

Small, cheap discoidal ceramic capacitors have existed since the 1930s, and remain in widespread use. Since the 1980s, surface mount packages for capacitors have been widely used. These packages are extremely small and lack connecting leads, allowing them to be soldered directly onto the surface of printed circuit boards. Surface mount components avoid undesirable high-frequency effects due to the leads and simplify automated assembly, although manual handling is made difficult due to their small size.

Mechanically controlled variable capacitors allow the plate spacing to be adjusted, for example by rotating or sliding a set of movable plates into alignment with a set of stationary plates. Low cost variable capacitors squeeze together alternating layers of aluminum and plastic with a screw. Electrical control of capacitance is achievable with varactors (or varicaps), which are reverse-biased semiconductor diodes whose depletion region width varies with applied voltage. They are used in phase-locked loops, amongst other applications.

Capacitor Markings

Most capacitors have numbers printed on their bodies to indicate their electrical characteristics. Larger capacitors like electrolytics usually display the actual capacitance together with the unit (for example, 220 µF). Smaller capacitors like ceramics, however, use a shorthand consisting of three numeric digits and a letter, where the digits indicate the capacitance in pF (calculated as XY

$\times$ 10^Z for digits XYZ) and the letter indicates the tolerance (J, K or M for ±5%, ±10% and ±20% respectively).

Additionally, the capacitor may show its working voltage, temperature and other relevant characteristics.

For typographical reasons, some manufacturers print "MF" on capacitors to indicate microfarads (μF).

Example

A capacitor with the text 473K 330V on its body has a capacitance of 47 $\times$ 10^3 pF = 47 nF (±10%) with a working voltage of 330 V. The working voltage of a capacitor is the highest voltage that can be applied across it without undue risk of breaking down the dielectric layer.

Historical

In the past, alternate capacitance subunits were used in historical electronic books; "mfd" and "mf" for microfarad (μF); "mmfd", "mmf", "$\mu\mu$F" for picofarad (pF); but are rarely used any more.

Applications

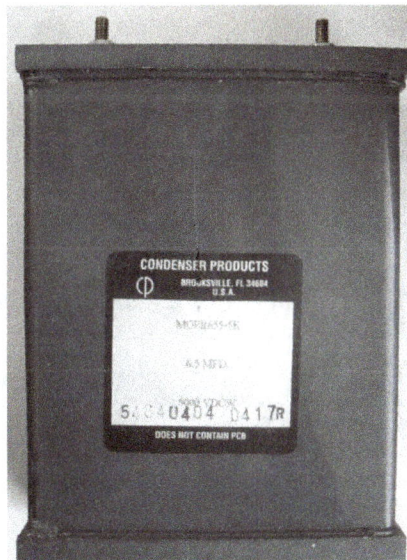

This mylar-film, oil-filled capacitor has very low inductance and low resistance, to provide the high-power (70 megawatt) and high speed (1.2 microsecond) discharge needed to operate a dye laser.

Energy Storage

A capacitor can store electric energy when disconnected from its charging circuit, so it can be used like a temporary battery, or like other types of rechargeable energy storage system. Capacitors are commonly used in electronic devices to maintain power supply while batteries are being changed. (This prevents loss of information in volatile memory.)

Conventional capacitors provide less than 360 joules per kilogram of specific energy, whereas a conventional alkaline battery has a density of 590 kJ/kg. There is an intermediate solution: Supercapacitors, which can accept and deliver charge much faster than batteries, and tolerate many more charge and discharge cycles than rechargeable batteries. They are however 10 times larger than conventional batteries for a given charge.

In car audio systems, large capacitors store energy for the amplifier to use on demand. Also for a flash tube a capacitor is used to hold the high voltage.

Pulsed Power and Weapons

Groups of large, specially constructed, low-inductance high-voltage capacitors (*capacitor banks*) are used to supply huge pulses of current for many pulsed power applications. These include electromagnetic forming, Marx generators, pulsed lasers (especially TEA lasers), pulse forming networks, radar, fusion research, and particle accelerators.

Large capacitor banks (reservoir) are used as energy sources for the exploding-bridgewire detonators or slapper detonators in nuclear weapons and other specialty weapons. Experimental work is under way using banks of capacitors as power sources for electromagnetic armour and electromagnetic railguns and coilguns.

Power Conditioning

A 10,000 microfarad capacitor in an amplifier power supply

Reservoir capacitors are used in power supplies where they smooth the output of a full or half wave rectifier. They can also be used in charge pump circuits as the energy storage element in the generation of higher voltages than the input voltage.

Capacitors are connected in parallel with the power circuits of most electronic devices and larger systems (such as factories) to shunt away and conceal current fluctuations from the primary power source to provide a "clean" power supply for signal or control circuits. Audio equipment, for example, uses several capacitors in this way, to shunt away power line hum before it gets into the signal circuitry. The capacitors act as a local reserve for the DC power source, and bypass AC currents from the power supply. This is used in car audio applications, when a stiffening capacitor compensates for the inductance and resistance of the leads to the lead-acid car battery.

Power Factor Correction

A high-voltage capacitor bank used for power factor correction on a power transmission system

In electric power distribution, capacitors are used for power factor correction. Such capacitors often come as three capacitors connected as a three phase load. Usually, the values of these capacitors are given not in farads but rather as a reactive power in volt-amperes reactive (var). The purpose is to counteract inductive loading from devices like electric motors and transmission lines to make the load appear to be mostly resistive. Individual motor or lamp loads may have capacitors for power factor correction, or larger sets of capacitors (usually with automatic switching devices) may be installed at a load center within a building or in a large utility substation.

Suppression and Coupling

Signal Coupling

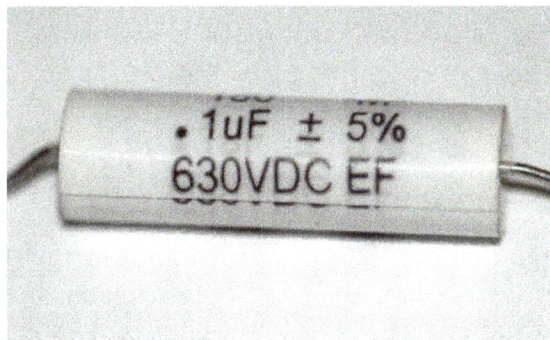

Polyester film capacitors are frequently used as coupling capacitors.

Because capacitors pass AC but block DC signals (when charged up to the applied dc voltage), they are often used to separate the AC and DC components of a signal. This method is known as *AC coupling* or "capacitive coupling". Here, a large value of capacitance, whose value need not be accurately controlled, but whose reactance is small at the signal frequency, is employed.

Decoupling

A decoupling capacitor is a capacitor used to protect one part of a circuit from the effect of another, for instance to suppress noise or transients. Noise caused by other circuit elements is shunted through the capacitor, reducing the effect they have on the rest of the circuit. It is most commonly used between the power supply and ground. An alternative name is *bypass capacitor* as it is used to bypass the power supply or other high impedance component of a circuit.

Decoupling capacitors need not always be discrete components. Capacitors used in these applications may be built into a printed circuit board, between the various layers. These are often referred to as embedded capacitors. The layers in the board contributing to the capacitive properties also function as power and ground planes, and have a dielectric in between them, enabling them to operate as a parallel plate capacitor.

High-pass and Low-pass Filters

Noise Suppression, Spikes, and Snubbers

When an inductive circuit is opened, the current through the inductance collapses quickly, creating a large voltage across the open circuit of the switch or relay. If the inductance is large enough, the energy will generate a spark, causing the contact points to oxidize, deteriorate, or sometimes weld together, or destroying a solid-state switch. A snubber capacitor across the newly opened circuit creates a path for this impulse to bypass the contact points, thereby preserving their life; these were commonly found in contact breaker ignition systems, for instance. Similarly, in smaller scale circuits, the spark may not be enough to damage the switch but will still radiate undesirable radio frequency interference (RFI), which a filter capacitor absorbs. Snubber capacitors are usually employed with a low-value resistor in series, to dissipate energy and minimize RFI. Such resistor-capacitor combinations are available in a single package.

Capacitors are also used in parallel to interrupt units of a high-voltage circuit breaker in order to equally distribute the voltage between these units. In this case they are called grading capacitors.

In schematic diagrams, a capacitor used primarily for DC charge storage is often drawn vertically in circuit diagrams with the lower, more negative, plate drawn as an arc. The straight plate indicates the positive terminal of the device, if it is polarized.

Motor Starters

In single phase squirrel cage motors, the primary winding within the motor housing is not capable of starting a rotational motion on the rotor, but is capable of sustaining one. To start the motor, a secondary "start" winding has a series non-polarized *starting capacitor* to introduce a lead in the sinusoidal current. When the secondary (start) winding is placed at an angle with respect to the primary (run) winding, a rotating electric field is created. The force of the rotational field is not constant, but is sufficient to start the rotor spinning. When the rotor comes close to operating speed, a centrifugal switch (or current-sensitive relay in series with the main winding) disconnects the capacitor. The start capacitor is typically mounted to the side of the motor housing. These are called capacitor-start motors, that have relatively high starting torque. Typically they can have up-to four times as much starting torque than a split-phase motor and are used on applications such as compressors, pressure washers and any small device requiring high starting torques.

Capacitor-run induction motors have a permanently connected phase-shifting capacitor in series with a second winding. The motor is much like a two-phase induction motor.

Motor-starting capacitors are typically non-polarized electrolytic types, while running capacitors are conventional paper or plastic film dielectric types.

Signal Processing

The energy stored in a capacitor can be used to represent information, either in binary form, as in DRAMs, or in analogue form, as in analog sampled filters and CCDs. Capacitors can be used in analog circuits as components of integrators or more complex filters and in negative feedback loop stabilization. Signal processing circuits also use capacitors to integrate a current signal.

Tuned Circuits

Capacitors and inductors are applied together in tuned circuits to select information in particular frequency bands. For example, radio receivers rely on variable capacitors to tune the station frequency. Speakers use passive analog crossovers, and analog equalizers use capacitors to select different audio bands.

The resonant frequency f of a tuned circuit is a function of the inductance (L) and capacitance (C) in series, and is given by:

$$f = \frac{1}{2\pi\sqrt{LC}}$$

where L is in henries and C is in farads.

Sensing

Most capacitors are designed to maintain a fixed physical structure. However, various factors can change the structure of the capacitor, and the resulting change in capacitance can be used to sense those factors.

Changing the dielectric:

> The effects of varying the characteristics of the **dielectric** can be used for sensing purposes. Capacitors with an exposed and porous dielectric can be used to measure humidity in air. Capacitors are used to accurately measure the fuel level in airplanes; as the fuel covers more of a pair of plates, the circuit capacitance increases. Squeezing the dielectric can change a capacitor at a few tens of bar pressure sufficiently that it can be used as a pressure sensor. A selected, but otherwise standard, polymer dielectric capacitor, when immersed in a compatible gas or liquid, can work usefully as a very low cost pressure sensor up to many hundreds of bar.

Changing the distance between the plates:

> Capacitors with a flexible plate can be used to measure strain or pressure. Industrial pressure transmitters used for process control use pressure-sensing diaphragms, which

form a capacitor plate of an oscillator circuit. Capacitors are used as the sensor in condenser microphones, where one plate is moved by air pressure, relative to the fixed position of the other plate. Some accelerometers use MEMS capacitors etched on a chip to measure the magnitude and direction of the acceleration vector. They are used to detect changes in acceleration, in tilt sensors, or to detect free fall, as sensors triggering airbag deployment, and in many other applications. Some fingerprint sensors use capacitors. Additionally, a user can adjust the pitch of a theremin musical instrument by moving their hand since this changes the effective capacitance between the user's hand and the antenna.

Changing the effective area of the plates:

Capacitive touch switches are now used on many consumer electronic products.

Oscillators

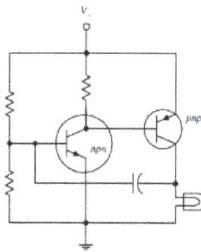

Example of a simple oscillator that requires a capacitor to function

A capacitor can possess spring-like qualities in an oscillator circuit. In the image example, a capacitor acts to influence the biasing voltage at the npn transistor's base. The resistance values of the voltage-divider resistors and the capacitance value of the capacitor together control the oscillatory frequency.

Producing Light

A light-emitting capacitor is made from a dielectric that uses phosphorescence to produce light. If one of the conductive plates is made with a transparent material, the light will be visible. Light-emitting capacitors are used in the construction of electroluminescent panels, for applications such as backlighting for laptop computers. In this case, the entire panel is a capacitor used for the purpose of generating light.

Hazards and Safety

The hazards posed by a capacitor are usually determined, foremost, by the amount of energy stored, which is the cause of things like electrical burns or heart fibrillation. Factors such as voltage and chassis material are of secondary consideration, which are more related to how easily a shock can be initiated rather than how much damage can occur.

Capacitors may retain a charge long after power is removed from a circuit; this charge can cause dangerous or even potentially fatal shocks or damage connected equipment. For example, even a seemingly innocuous device such as a disposable-camera flash unit, powered by a 1.5 volt AA battery, has a capacitor which may contain over 15 joules of energy and be charged to over

300 volts. This is easily capable of delivering a shock. Service procedures for electronic devices usually include instructions to discharge large or high-voltage capacitors, for instance using a Brinkley stick. Capacitors may also have built-in discharge resistors to dissipate stored energy to a safe level within a few seconds after power is removed. High-voltage capacitors are stored with the terminals shorted, as protection from potentially dangerous voltages due to dielectric absorption or from transient voltages the capacitor may pick up from static charges or passing weather events.

Some old, large oil-filled paper or plastic film capacitors contain polychlorinated biphenyls (PCBs). It is known that waste PCBs can leak into groundwater under landfills. Capacitors containing PCB were labelled as containing "Askarel" and several other trade names. PCB-filled paper capacitors are found in very old (pre-1975) fluorescent lamp ballasts, and other applications.

Capacitors may catastrophically fail when subjected to voltages or currents beyond their rating, or as they reach their normal end of life. Dielectric or metal interconnection failures may create arcing that vaporizes the dielectric fluid, resulting in case bulging, rupture, or even an explosion. Capacitors used in RF or sustained high-current applications can overheat, especially in the center of the capacitor rolls. Capacitors used within high-energy capacitor banks can violently explode when a short in one capacitor causes sudden dumping of energy stored in the rest of the bank into the failing unit. High voltage vacuum capacitors can generate soft X-rays even during normal operation. Proper containment, fusing, and preventive maintenance can help to minimize these hazards.

High-voltage capacitors can benefit from a pre-charge to limit in-rush currents at power-up of high voltage direct current (HVDC) circuits. This will extend the life of the component and may mitigate high-voltage hazards.

Swollen caps of electrolytic capacitors – special design of semi-cut caps prevents capacitors from bursting

This high-energy capacitor from a defibrillator can deliver over 500 joules of energy.
A resistor is connected between the terminals for safety, to allow the stored energy to be released.

Catastrophic failure

Inductor

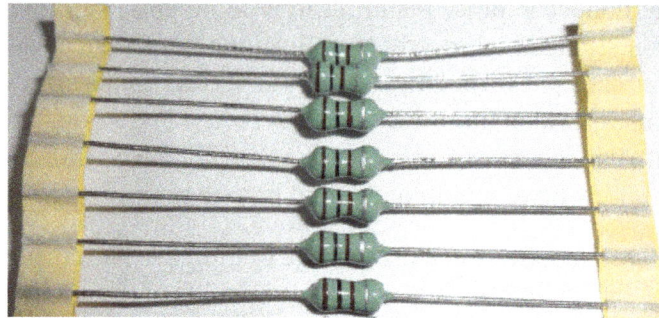

Axial lead inductors (100 µH)

An inductor, also called a coil or reactor, is a passive two-terminal electrical component which resists changes in electric current passing through it. It consists of a conductor such as a wire, usually wound into a coil. Energy is stored in a magnetic field in the coil as long as current flows. When the current flowing through an inductor changes, the time-varying magnetic field induces a voltage in the conductor, according to Faraday's law of electromagnetic induction. According to Lenz's law the direction of induced electromotive force (or "e.m.f.") is always such that it opposes the change in current that created it. As a result, inductors always oppose a change in current, in the same way that a flywheel opposes a change in rotational velocity. Care should be taken not to confuse this with the resistance provided by a resistor.

An inductor is characterized by its *inductance*, the ratio of the voltage to the rate of change of current, which has units of henries (H). Inductors have values that typically range from 1 µH (10^{-6}H) to 1 H. Many inductors have a magnetic core made of iron or ferrite inside the coil, which serves to increase the magnetic field and thus the inductance. Along with capacitors and resistors, inductors are one of the three passive linear circuit elements that make up electric circuits. Inductors are widely used in alternating current (AC) electronic equipment, particularly in radio equipment. They are used to block AC while allowing DC to pass; inductors designed for this purpose are called chokes. They are also used in electronic filters to separate signals of different frequencies, and in combination with capacitors to make tuned circuits, used to tune radio and TV receivers.

Overview

Inductance (L) results from the magnetic field around a current-carrying conductor; the electric current through the conductor creates a magnetic flux. Mathematically speaking, inductance is determined by how much magnetic flux φ through the circuit is created by a given current i

$$L = \frac{\phi}{i} \qquad (1)$$

Inductors that have ferromagnetic cores are nonlinear; the inductance changes with the current, in this more general case inductance is defined as

$$L = \frac{d\phi}{di}$$

Any wire or other conductor will generate a magnetic field when current flows through it, so every conductor has some inductance. The inductance of a circuit depends on the geometry of the current path as well as the magnetic permeability of nearby materials. An inductor is a component consisting of a wire or other conductor shaped to increase the magnetic flux through the circuit, usually in the shape of a coil or helix. Winding the wire into a coil increases the number of times the magnetic flux lines link the circuit, increasing the field and thus the inductance. The more turns, the higher the inductance. The inductance also depends on the shape of the coil, separation of the turns, and many other factors. By adding a "magnetic core" made of a ferromagnetic material like iron inside the coil, the magnetizing field from the coil will induce magnetization in the material, increasing the magnetic flux. The high permeability of a ferromagnetic core can increase the inductance of a coil by a factor of several thousand over what it would be without it.

Constitutive Equation

Any change in the current through an inductor creates a changing flux, inducing a voltage across the inductor. By Faraday's law of induction, the voltage induced by any change in magnetic flux through the circuit is

$$v = \frac{d\phi}{dt}$$

From (1) above

$$v = \frac{d}{dt}(Li) = L\frac{di}{dt} \qquad (2)$$

So inductance is also a measure of the amount of electromotive force (voltage) generated for a given rate of change of current. For example, an inductor with an inductance of 1 henry produces an EMF of 1 volt when the current through the inductor changes at the rate of 1 ampere per second. This is usually taken to be the constitutive relation (defining equation) of the inductor.

The dual of the inductor is the capacitor, which stores energy in an electric field rather than a magnetic field. Its current-voltage relation is obtained by exchanging current and voltage in the inductor equations and replacing L with the capacitance C.

Lenz's Law

The polarity (direction) of the induced voltage is given by Lenz's law, which states that it will be such as to oppose the change in current. For example, if the current through an inductor is increasing, the induced voltage will be positive at the terminal through which the current enters and negative at the terminal through which it leaves, tending to oppose the additional current. The energy from the external circuit necessary to overcome this potential "hill" is being stored in the magnetic field of the inductor; the inductor is said to be "charging" or "energizing". If the current is decreasing, the induced voltage will be negative at the terminal through which the current enters and positive at the terminal through which it leaves, tending to maintain the current. Energy from the magnetic field is being returned to the circuit; the inductor is said to be "discharging".

Ideal and Real Inductors

In circuit theory, inductors are idealized as obeying the mathematical relation (2) above precisely. An "ideal inductor" has inductance, but no resistance or capacitance, and does not dissipate or radiate energy. However real inductors have side effects which cause their behavior to depart from this simple model. They have resistance (due to the resistance of the wire and energy losses in core material), and parasitic capacitance (due to the electric field between the turns of wire which are at slightly different potentials). At high frequencies the capacitance begins to affect the inductor's behavior; at some frequency, real inductors behave as resonant circuits, becoming self-resonant. Above the resonant frequency the capacitive reactance becomes the dominant part of the impedance. At higher frequencies, resistive losses in the windings increase due to skin effect and proximity effect.

Inductors with ferromagnetic cores have additional energy losses due to hysteresis and eddy currents in the core, which increase with frequency. At high currents, iron core inductors also show gradual departure from ideal behavior due to nonlinearity caused by magnetic saturation of the core. An inductor may radiate electromagnetic energy into surrounding space and circuits, and may absorb electromagnetic emissions from other circuits, causing electromagnetic interference (EMI). Real-world inductor applications may consider these parasitic parameters as important as the inductance.

Applications

Example of signal filtering. In this configuration, the inductor blocks AC current, while allowing DC current to pass.

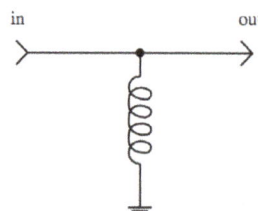

Example of signal filtering. In this configuration, the inductor decouples DC current, while allowing AC current to pass.

Large 50 MVAR three-phase iron-core loading inductor at an Austrian utility substation

A ferrite "bead" choke, consisting of an encircling ferrite cylinder, removes electronic noise from a computer power cord.

Inductors are used extensively in analog circuits and signal processing. Applications range from the use of large inductors in power supplies, which in conjunction with filter capacitors remove residual hums known as the mains hum or other fluctuations from the direct current output, to the small inductance of the ferrite bead or torus installed around a cable to prevent radio frequency interference from being transmitted down the wire. Inductors are used as the energy storage device in many switched-mode power supplies to produce DC current. The inductor supplies energy to the circuit to keep current flowing during the "off" switching periods.

An inductor connected to a capacitor forms a tuned circuit, which acts as a resonator for oscillating current. Tuned circuits are widely used in radio frequency equipment such as radio transmitters and receivers, as narrow bandpass filters to select a single frequency from a composite signal, and in electronic oscillators to generate sinusoidal signals.

Two (or more) inductors in proximity that have coupled magnetic flux (mutual inductance) form a transformer, which is a fundamental component of every electric utility power grid. The efficiency of a transformer may decrease as the frequency increases due to eddy currents in the core material and skin effect on the windings. The size of the core can be decreased at higher frequencies. For this reason, aircraft use 400 hertz alternating current rather than the usual 50 or 60 hertz, allowing a great saving in weight from the use of smaller transformers.

Inductors are also employed in electrical transmission systems, where they are used to limit switching currents and fault currents. In this field, they are more commonly referred to as reactors.

Because inductors have complicated side effects (detailed below) which cause them to depart from ideal behavior, because they can radiate electromagnetic interference (EMI), and most of all because of their bulk which prevents them from being integrated on semiconductor chips, the use of inductors is declining in modern electronic devices, particularly compact portable devices. Real inductors are increasingly being replaced by active circuits such as the gyrator which can synthesize inductance using capacitors.

Inductor Construction

A ferrite core inductor with two 47 mH windings.

An inductor usually consists of a coil of conducting material, typically insulated copper wire, wrapped around a core either of plastic or of a ferromagnetic (or ferrimagnetic) material; the latter is called an "iron core" inductor. The high permeability of the ferromagnetic core increases the magnetic field and confines it closely to the inductor, thereby increasing the inductance. Low frequency inductors are constructed like transformers, with cores of electrical steel laminated to prevent eddy currents. 'Soft' ferrites are widely used for cores above audio frequencies, since they do not cause the large energy losses at high frequencies that ordinary iron alloys do. Inductors come in many shapes. Most are constructed as enamel coated wire (magnet wire) wrapped around a ferrite bobbin with wire exposed on the outside, while some enclose the wire completely in ferrite and are referred to as "shielded". Some inductors have an adjustable core, which enables changing of the inductance. Inductors used to block very high frequencies are sometimes made by stringing a ferrite bead on a wire.

Small inductors can be etched directly onto a printed circuit board by laying out the trace in a spiral pattern. Some such planar inductors use a planar core.

Small value inductors can also be built on integrated circuits using the same processes that are used to make transistors. Aluminium interconnect is typically used, laid out in a spiral coil pattern. However, the small dimensions limit the inductance, and it is far more common to use a circuit called a "gyrator" that uses a capacitor and active components to behave similarly to an inductor.

Types of Inductor

Air core Inductor

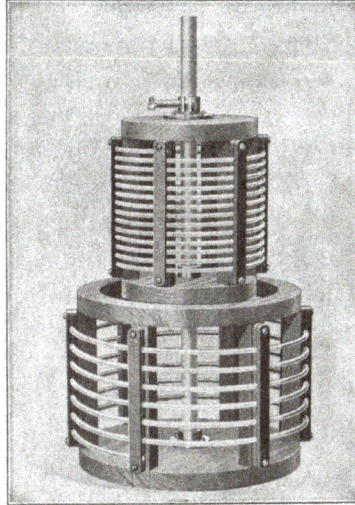

Resonant oscillation transformer from a spark gap transmitter. Coupling can be adjusted by moving the top coil on the support rod. Shows high Q construction with spaced turns of large diameter tubing.

The term *air core coil* describes an inductor that does not use a magnetic core made of a ferromagnetic material. The term refers to coils wound on plastic, ceramic, or other nonmagnetic forms, as well as those that have only air inside the windings. Air core coils have lower inductance than ferromagnetic core coils, but are often used at high frequencies because they are free from energy losses called core losses that occur in ferromagnetic cores, which increase with frequency. A side effect that can occur in air core coils in which the winding is not rigidly supported on a form is 'microphony': mechanical vibration of the windings can cause variations in the inductance.

Radio Frequency Inductor

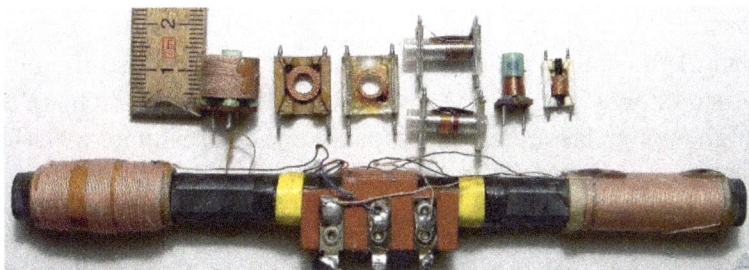

Collection of RF inductors, showing techniques to reduce losses. The three top left and the ferrite loopstick or rod antenna, bottom, have basket windings.

At high frequencies, particularly radio frequencies (RF), inductors have higher resistance and other losses. In addition to causing power loss, in resonant circuits this can reduce the Q factor of the circuit, broadening the bandwidth. In RF inductors, which are mostly air core types, specialized construction techniques are used to minimize these losses. The losses are due to these effects:

- Skin effect: The resistance of a wire to high frequency current is higher than its resistance to direct current because of skin effect. Radio frequency alternating current does not penetrate far into the body of a conductor but travels along its surface. Therefore, in a solid wire, most of the cross sectional area of the wire is not used to conduct the current, which is in a narrow annulus on the surface. This effect increases the resistance of the wire in the coil, which may already have a relatively high resistance due to its length and small diameter.

- Proximity effect: Another similar effect that also increases the resistance of the wire at high frequencies is proximity effect, which occurs in parallel wires that lie close to each other. The individual magnetic field of adjacent turns induces eddy currents in the wire of the coil, which causes the current in the conductor to be concentrated in a thin strip on the side near the adjacent wire. Like skin effect, this reduces the effective cross-sectional area of the wire conducting current, increasing its resistance.

High Q tank coil in a shortwave transmitter   *(left)* Spiderweb coil *(right)* Adjustable ferrite slug-tuned RF coil
                                                                                           with basketweave winding and litz wire

- Dielectric losses: The high frequency electric field near the conductors in a tank coil can cause the motion of polar molecules in nearby insulating materials, dissipating energy as heat. So coils used for tuned circuits are often not wound on coil forms but are suspended in air, supported by narrow plastic or ceramic strips.

- Parasitic capacitance: The capacitance between individual wire turns of the coil, called parasitic capacitance, does not cause energy losses but can change the behavior of the coil. Each turn of the coil is at a slightly different potential, so the electric field between neighboring turns stores charge on the wire, so the coil acts as if it has a capacitor in parallel with it. At a high enough frequency this capacitance can resonate with the inductance of the coil forming a tuned circuit, causing the coil to become self-resonant.

To reduce parasitic capacitance and proximity effect, high Q RF coils are constructed to avoid having many turns lying close together, parallel to one another. The windings of RF coils are often limited to a single layer, and the turns are spaced apart. To reduce resistance due to skin effect, in high-power inductors such as those used in transmitters the windings are sometimes made of a metal strip or tubing which has a larger surface area, and the surface is silver-plated.

- Basket-weave coils: To reduce proximity effect and parasitic capacitance, multilayer RF coils are wound in patterns in which successive turns are not parallel but crisscrossed at an angle; these are often called *honeycomb* or *basket-weave* coils. These are occasionally wound on a vertical insulating supports with dowels or slots, with the wire weaving in and out through the slots.

- Spiderweb coils: Another construction technique with similar advantages is flat spiral coils. These are often wound on a flat insulating support with radial spokes or slots, with the wire weaving in and out through the slots; these are called *spiderweb* coils. The form has an odd number of slots, so successive turns of the spiral lie on opposite sides of the form, increasing separation.

- Litz wire: To reduce skin effect losses, some coils are wound with a special type of radio frequency wire called litz wire. Instead of a single solid conductor, litz wire consists of a number of smaller wire strands that carry the current. Unlike ordinary stranded wire, the strands are insulated from each other, to prevent skin effect from forcing the current to the surface, and are twisted or braided together. The twist pattern ensures that each wire strand spends the same amount of its length on the outside of the wire bundle, so skin effect distributes the current equally between the strands, resulting in a larger cross-sectional conduction area than an equivalent single wire.

Ferromagnetic Core Inductor

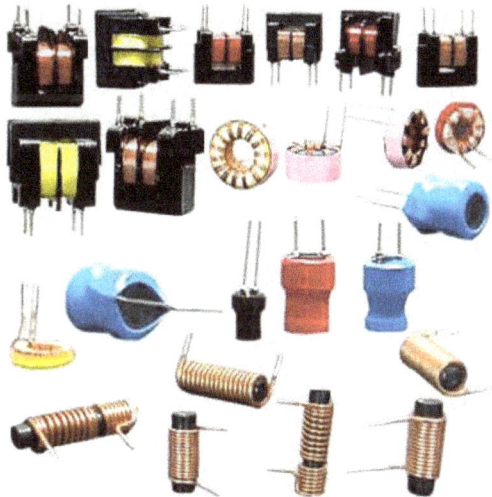

A variety of types of ferrite core inductors and transformers

Ferromagnetic-core or iron-core inductors use a magnetic core made of a ferromagnetic or ferrimagnetic material such as iron or ferrite to increase the inductance. A magnetic core can increase the inductance of a coil by a factor of several thousand, by increasing the magnetic field due to its higher magnetic permeability. However the magnetic properties of the core material cause several side effects which alter the behavior of the inductor and require special construction:

- Core losses: A time-varying current in a ferromagnetic inductor, which causes a time-varying magnetic field in its core, causes energy losses in the core material that are dissipated as heat, due to two processes:

  o Eddy currents: From Faraday's law of induction, the changing magnetic field can induce circulating loops of electric current in the conductive metal core. The energy in these currents is dissipated as heat in the resistance of the core material. The amount of energy lost increases with the area inside the loop of current.

o Hysteresis: Changing or reversing the magnetic field in the core also causes losses due to the motion of the tiny magnetic domains it is composed of. The energy loss is proportional to the area of the hysteresis loop in the BH graph of the core material. Materials with low coercivity have narrow hysteresis loops and so low hysteresis losses.

For both of these processes, the energy loss per cycle of alternating current is constant, so core losses increase linearly with frequency. Online core loss calculators are available to calculate the energy loss. Using inputs such as input voltage, output voltage, output current, frequency, ambient temperature, and inductance these calculators can predict the losses of the inductors core and AC/DC based on the operating condition of the circuit being used.

• Nonlinearity: If the current through a ferromagnetic core coil is high enough that the magnetic core saturates, the inductance will not remain constant but will change with the current through the device. This is called nonlinearity and results in distortion of the signal. For example, audio signals can suffer intermodulation distortion in saturated inductors. To prevent this, in linear circuits the current through iron core inductors must be limited below the saturation level. Some laminated cores have a narrow air gap in them for this purpose, and powdered iron cores have a distributed air gap. This allows higher levels of magnetic flux and thus higher currents through the inductor before it saturates.

Laminated Core Inductor

Laminated iron core ballast inductor for a metal halide lamp

Low-frequency inductors are often made with laminated cores to prevent eddy currents, using construction similar to transformers. The core is made of stacks of thin steel sheets or laminations oriented parallel to the field, with an insulating coating on the surface. The insulation prevents eddy currents between the sheets, so any remaining currents must be within the cross sectional area of the individual laminations, reducing the area of the loop and thus reducing the energy losses greatly. The laminations are made of low-coercivity silicon steel, to reduce hysteresis losses.

Ferrite-core Inductor

For higher frequencies, inductors are made with cores of ferrite. Ferrite is a ceramic ferrimagnetic material that is nonconductive, so eddy currents cannot flow within it. The formulation of ferrite is $xxFe_2O_4$ where xx represents various metals. For inductor cores soft ferrites are used, which have low coercivity and thus low hysteresis losses. Another similar material is powdered iron cemented with a binder.

Toroidal Core Inductor

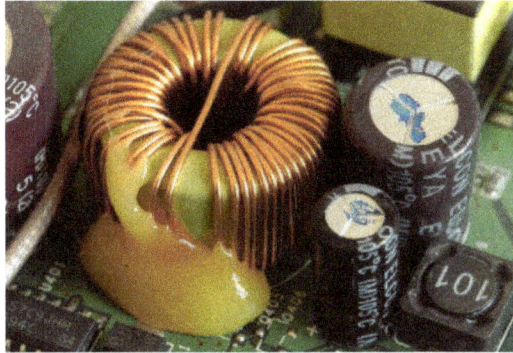

Toroidal inductor in the power supply of a wireless router

In an inductor wound on a straight rod-shaped core, the magnetic field lines emerging from one end of the core must pass through the air to re-enter the core at the other end. This reduces the field, because much of the magnetic field path is in air rather than the higher permeability core material. A higher magnetic field and inductance can be achieved by forming the core in a closed magnetic circuit. The magnetic field lines form closed loops within the core without leaving the core material. The shape often used is a toroidal or doughnut-shaped ferrite core. Because of their symmetry, toroidal cores allow a minimum of the magnetic flux to escape outside the core (called *leakage flux*), so they radiate less electromagnetic interference than other shapes. Toroidal core coils are manufactured of various materials, primarily ferrite, powdered iron and laminated cores.

Choke

An MF or HF radio choke for tenths of an ampere, and a ferrite bead VHF choke for several amperes.

A choke is designed specifically for blocking higher-frequency alternating current (AC) in an electrical circuit, while allowing lower frequency or DC current to pass. It usually consists of a coil of insulated wire often wound on a magnetic core, although some consist of a donut-shaped "bead" of ferrite material strung on a wire. Like other inductors, chokes resist changes to the current passing through them, and so alternating currents of higher frequency, which reverse direction rapidly, are resisted more than currents of lower frequency; the choke's impedance increases with frequency. Its low electrical resistance allows both AC and DC to pass with little power loss, but it can limit the amount of AC passing through it due to its reactance.

Variable Inductor

(left) Inductor with a threaded ferrite slug *(visible at top)* that can be turned to move it into or out of the coil. 4.2 cm high. *(right)* A variometer used in radio receivers in the 1920s

A "roller coil", an adjustable air-core RF inductor used in the tuned circuits of radio transmitters. One of the contacts to the coil is made by the small grooved wheel, which rides on the wire. Turning the shaft rotates the coil, moving the contact wheel up or down the coil, allowing more or fewer turns of the coil into the circuit, to change the inductance.

Probably the most common type of variable inductor today is one with a moveable ferrite magnetic core, which can be slid or screwed in or out of the coil. Moving the core farther into the coil increases the permeability, increasing the magnetic field and the inductance. Many inductors used in radio applications (usually less than 100 MHz) use adjustable cores in order to tune such inductors to their desired value, since manufacturing processes have certain tolerances (inaccuracy). Sometimes such cores for frequencies above 100 MHz are made from highly conductive non-magnetic material such as aluminum. They decrease the inductance because the magnetic field must bypass them.

Air core inductors can use sliding contacts or multiple taps to increase or decrease the number of turns included in the circuit, to change the inductance. A type much used in the past but mostly obsolete today has a spring contact that can slide along the bare surface of the windings. The disadvantage of this type is that the contact usually short-circuits one or more turns. These turns act like a single-turn short-circuited transformer secondary winding; the large currents induced in them cause power losses.

A type of continuously variable air core inductor is the *variometer*. This consists of two coils with the same number of turns connected in series, one inside the other. The inner coil is mounted on a shaft so its axis can be turned with respect to the outer coil. When the two coils' axes are collinear, with the magnetic fields pointing in the same direction, the fields add and the inductance is maximum. When the inner coil is turned so its axis is at an angle with the outer, the mutual inductance between them is smaller so the total inductance is less. When the inner coil is turned 180° so the coils are collinear with their magnetic fields opposing, the two fields cancel each other and the inductance is very small. This type has the advantage that it is continuously variable over a wide range. It is used in antenna tuners and matching circuits to match low frequency transmitters to their antennas.

Another method to control the inductance without any moving parts requires an additional DC current bias winding which controls the permeability of an easily saturable core material.

Circuit Theory

The effect of an inductor in a circuit is to oppose changes in current through it by developing a voltage across it proportional to the rate of change of the current. An ideal inductor would offer no resistance to a constant direct current; however, only superconducting inductors have truly zero electrical resistance.

The relationship between the time-varying voltage $v(t)$ across an inductor with inductance L and the time-varying current $i(t)$ passing through it is described by the differential equation:

$$v(t) = L\frac{di(t)}{dt}$$

When there is a sinusoidal alternating current (AC) through an inductor, a sinusoidal voltage is induced. The amplitude of the voltage is proportional to the product of the amplitude (I_p) of the current and the frequency (f) of the current.

$$i(t) = I_\mathrm{P} \sin(2\pi ft)$$

$$\frac{di(t)}{dt} = 2\pi fI_\mathrm{P} \cos(2\pi ft)$$

$$v(t) = 2\pi fLI_\mathrm{P} \cos(2\pi ft)$$

In this situation, the phase of the current lags that of the voltage by $\pi/2$ (90°). For sinusoids, as the voltage across the inductor goes to its maximum value, the current goes to zero, and as the voltage across the inductor goes to zero, the current through it goes to its maximum value.

If an inductor is connected to a direct current source with value I via a resistance R, and then the current source is short-circuited, the differential relationship above shows that the current through the inductor will discharge with an exponential decay:

$$i(t) = Ie^{-\frac{R}{L}t}$$

Reactance

The ratio of the peak voltage to the peak current in an inductor energised from a sinusoidal source is called the reactance and is denoted X_L. The subscript is to distinguish inductive reactance from capacitive reactance due to capacitance.

$$X_\mathrm{L} = \frac{V_\mathrm{P}}{I_\mathrm{P}} = \frac{2\pi fLI_\mathrm{P}}{I_\mathrm{P}}$$

Thus,

$$X_\mathrm{L} = 2\pi fL$$

Reactance is measured in the same units as resistance (ohms) but is not actually a resistance. A resistor will dissipate energy as heat when a current passes. This does not happen with an inductor; rather, energy is stored in the magnetic field as the current builds and later returned to the circuit as the current falls. Inductive reactance is strongly frequency dependent. At low frequency the reactance falls, and for a steady current (zero frequency) the inductor behaves as a short-circuit. At increasing frequency, on the other hand, the reactance increases and at a sufficiently high frequency the inductor approaches an open circuit.

Laplace Circuit Analysis (s-domain)

When using the Laplace transform in circuit analysis, the impedance of an ideal inductor with no initial current is represented in the s domain by:

$$Z(s) = Ls$$

where

L is the inductance, and

s is the complex frequency.

If the inductor does have initial current, it can be represented by:

- adding a voltage source in series with the inductor, having the value:

$$LI_0$$

where

L is the inductance, and

I_0 is the initial current in the inductor.

(*Note that the source should have a polarity that is aligned with the initial current*)

- or by adding a current source in parallel with the inductor, having the value:

$$\frac{I_0}{s}$$

where

I_0 is the initial current in the inductor.

s is the complex frequency.

Inductor Networks

Inductors in a parallel configuration each have the same potential difference (voltage). To find their total equivalent inductance (L_{eq}):

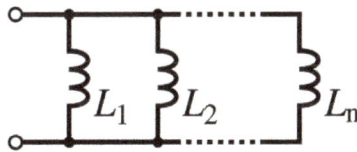

$$\frac{1}{L_{eq}} = \frac{1}{L_1} + \frac{1}{L_2} + \cdots + \frac{1}{L_n}$$

The current through inductors in series stays the same, but the voltage across each inductor can be different. The sum of the potential differences (voltage) is equal to the total voltage. To find their total inductance:

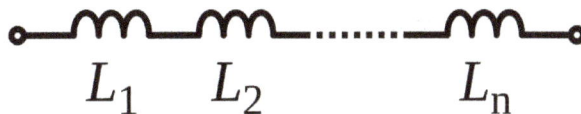

$$L_{eq} = L_1 + L_2 + \cdots + L_n$$

These simple relationships hold true only when there is no mutual coupling of magnetic fields between individual inductors.

Stored Energy

Neglecting losses, the energy (measured in joules, in SI) stored by an inductor is equal to the amount of work required to establish the current through the inductor, and therefore the magnetic field. This is given by:

$$E_{stored} = \frac{1}{2}LI^2$$

where L is inductance and I is the current through the inductor.

This relationship is only valid for linear (non-saturated) regions of the magnetic flux linkage and current relationship. In general if one decides to find the energy stored in a LTI inductor that has initial current in a specific time between t_0 and t_1 can use this:

$$E = \int_{t_0}^{t_1} P(t)dt = \frac{1}{2}LI(t_1)^2 - \frac{1}{2}LI(t_0)^2$$

Q Factor

An ideal inductor would have no resistance or energy losses. However, real inductors have winding resistance from the metal wire forming the coils. Since the winding resistance appears as a resistance in series with the inductor, it is often called the *series resistance*. The inductor's series resistance converts electric current through the coils into heat, thus causing a loss of inductive quality. The quality factor (or Q) of an inductor is the ratio of its inductive reactance to its resistance at a given frequency, and is a measure of its efficiency. The higher the Q factor of the inductor, the closer it approaches the behavior of an ideal, lossless, inductor. High Q inductors are used with capacitors to make resonant circuits in radio transmitters and receivers. The higher the Q is, the narrower the bandwidth of the resonant circuit.

The Q factor of an inductor can be found through the following formula, where L is the inductance, R is the inductor's effective series resistance, ω is the radian operating frequency, and the product ωL is the inductive reactance:

$$Q = \frac{\omega L}{R}$$

Notice that Q increases linearly with frequency if L and R are constant. Although they are constant at low frequencies, the parameters vary with frequency. For example, skin effect, proximity effect, and core losses increase R with frequency; winding capacitance and variations in permeability with frequency affect L.

Qualitatively, at low frequencies and within limits, increasing the number of turns N improves Q because L varies as N^2 while R varies linearly with N. Similarly, increasing the radius r of an inductor improves Q because L varies as r^2 while R varies linearly with r. So high Q air core inductors often have large diameters and many turns. Both of those examples assume the diameter of the wire stays the same, so both examples use proportionally more wire (copper). If the total mass of wire is held constant, then there would be no advantage to increasing the number of turns or the radius of the turns because the wire would have to be proportionally thinner.

Using a high permeability ferromagnetic core can greatly increase the inductance for the same amount of copper, so the core can also increase the Q. Cores however also introduce losses that increase with frequency. The core material is chosen for best results for the frequency band. At VHF or higher frequencies an air core is likely to be used.

Inductors wound around a ferromagnetic core may saturate at high currents, causing a dramatic decrease in inductance (and Q). This phenomenon can be avoided by using a (physically larger) air core inductor. A well designed air core inductor may have a Q of several hundred.

Inductance Formulas

The table below lists some common simplified formulas for calculating the approximate inductance of several inductor constructions.

Construction	Formula	Notes
Cylindrical air-core coil	$$L = \frac{1}{l}\mu_0 K N^2 A$$ • L = inductance in henries (H) • μ_0 = permeability of free space = 4×10^{-7} H/m • K = Nagaoka coefficient • N = number of turns • A = area of cross-section of the coil in square metres (m²) • l = length of coil in metres (m)	The exact calculation of K is complex. K is approximately unity for a coil which is much longer than its diameter and is tightly wound using small gauge wire (so that it approximates a current sheet).
Straight wire conductor	$$L = \frac{\mu_0}{2\pi}\left(l \ln\left[\frac{1}{c}\left(l + \sqrt{l^2 + c^2}\right)\right] - \sqrt{l^2 + c^2} + c + \frac{l}{4 + c\sqrt{\frac{2}{p}\omega\mu}}\right)$$ • L = inductance • l = cylinder length • c = cylinder radius • μ_0 = permeability of free space = 4×10^{-7} H/m • μ = conductor permeability • p = resistivity • ω = phase rate	Exact if $\omega = 0$ or $\omega = \infty$

	$$L = \frac{1}{5}l\left[\ln\left(\frac{4l}{d}\right) - 1\right]$$ (when $d^2 f \gg 1$ mm² MHz) $$L = \frac{1}{5}l\left[\ln\left(\frac{4l}{d}\right) - \frac{3}{4}\right]$$ ((when $d^2 f \ll 1$ mm² MHz) • L = inductance (nH) • l = length of conductor (mm) • d = diameter of conductor (mm) • f = frequency	• Cu or Al (i.e., relative permeability is one) • $l > 100\,d$
Short air-core cylindrical coil	$$L = \frac{r^2 N^2}{9r + 10l}$$ • L = inductance (µH) • r = outer radius of coil (in) • l = length of coil (in) • N = number of turns	
Multilayer air-core coil	$$L = \frac{4}{5} \cdot \frac{r^2 N^2}{6r + 9l + 10d}$$ • L = inductance (µH) • r = mean radius of coil (in) • l = physical length of coil winding (in) • N = number of turns • d = depth of coil (outer radius minus inner radius) (in)	
Flat spiral air-core coil	$$L = \frac{r^2 N^2}{20r + 28d}$$ • L = inductance (µH) • r = mean radius of coil (cm) • N = number of turns • d = depth of coil (outer radius minus inner radius) (cm)	accurate to within 5 percent for $d > 0.2\,r$.
	$$L = \frac{r^2 N^2}{8r + 11d}$$ • L = inductance (µH) • r = mean radius of coil (in) • N = number of turns • d = depth of coil (outer radius minus inner radius) (in)	

Toroidal core (circular cross-section)	$$L = 0.01595N^2\left(D - \sqrt{D^2 - d^2}\right)$$ • L = inductance (µH) • d = diameter of coil winding (in) • N = number of turns • D = 2 * radius of revolution (in)	
	$$L \approx 0.007975\frac{d^2 N^2}{D}$$ • L = inductance (µH) • d = diameter of coil winding (in) • N = number of turns • D = 2 * radius of revolution (in)	approximation when $d < 0.1\,D$
Toroidal core (rectangular cross-section)	$$L = 0.00508N^2 h \ln\left(\frac{d_2}{d_1}\right)$$ • L = inductance (µH) • d_1 = inside diameter of toroid (in) • d_2 = outside diameter of toroid (in) • N = number of turns • h = height of toroid (in)	

Diode

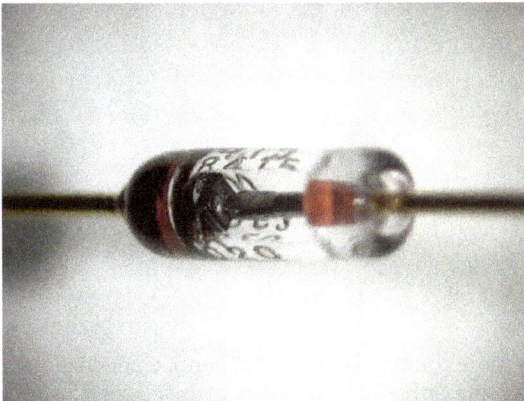

Closeup of a diode, showing the square-shaped semiconductor crystal *(black object on left)*.

Extreme macro photo of a Chinese diode of the seventies.

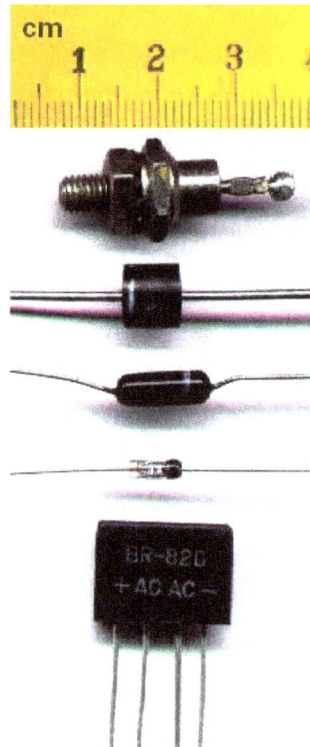

Various semiconductor diodes. Bottom: A bridge rectifier. In most diodes, a white or black painted band identifies the cathode into which electrons will flow when the diode is conducting. Electron flow is the reverse of conventional current flow.

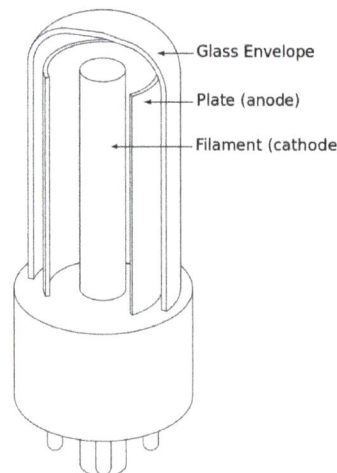

Structure of a vacuum tube diode. The filament may be bare, or more commonly (as shown here), embedded within and insulated from an enclosing cathode.

In electronics, a diode is a two-terminal electronic component that conducts primarily in one direction (asymmetric conductance); it has low (ideally zero) resistance to the flow of current in one direction, and high (ideally infinite) resistance in the other. A semiconductor diode, the most common type today, is a crystalline piece of semiconductor material with a p–n junction connected to two electrical terminals. A vacuum tube diode has two electrodes, a plate (anode) and a heated cathode. Semiconductor diodes were the first semiconductor electronic devices. The discovery of crystals' rectifying abilities was made by German physicist Ferdinand Braun in 1874. The first semiconductor diodes, called cat's whisker diodes, developed around 1906, were made of mineral crystals such as galena. Today, most diodes are made of silicon, but other semiconductors such as selenium or germanium are sometimes used.

Main Functions

The most common function of a diode is to allow an electric current to pass in one direction (called the diode's *forward* direction), while blocking current in the opposite direction (the *reverse* direction). Thus, the diode can be viewed as an electronic version of a check valve. This unidirectional behavior is called rectification, and is used to convert alternating current (AC) to direct current (DC), including extraction of modulation from radio signals in radio receivers—these diodes are forms of rectifiers.

However, diodes can have more complicated behavior than this simple on–off action, because of their nonlinear current-voltage characteristics. Semiconductor diodes begin conducting electricity only if a certain threshold voltage or cut-in voltage is present in the forward direction (a state in which the diode is said to be *forward-biased*). The voltage drop across a forward-biased diode varies only a little with the current, and is a function of temperature; this effect can be used as a temperature sensor or as a voltage reference.

A semiconductor diode's current–voltage characteristic can be tailored by selecting the semiconductor materials and the doping impurities introduced into the materials during manufacture. These techniques are used to create special-purpose diodes that perform many different functions. For example, diodes are used to regulate voltage (Zener diodes), to protect circuits from high voltage surges (avalanche diodes), to electronically tune radio and TV receivers (varactor diodes), to generate radio-frequency oscillations (tunnel diodes, Gunn diodes, IMPATT diodes), and to produce light (light-emitting diodes). Tunnel, Gunn and IMPATT diodes exhibit negative resistance, which is useful in microwave and switching circuits.

Diodes, both vacuum and semiconductor, can be used as shot-noise generators.

History

Thermionic (vacuum tube) diodes and solid state (semiconductor) diodes were developed separately, at approximately the same time, in the early 1900s, as radio receiver detectors. Until the 1950s vacuum tube diodes were used more frequently in radios because the early point-contact type semiconductor diodes were less stable. In addition, most receiving sets had vacuum tubes for amplification that could easily have the thermionic diodes included in the tube (for example the 12SQ7 double diode triode), and vacuum tube rectifiers and gas-filled rectifiers were capable of handling some high voltage/high current rectification tasks better than the semiconductor diodes (such as selenium rectifiers) which were available at that time.

Vacuum Tube Diodes

In 1873, Frederick Guthrie discovered the basic principle of operation of thermionic diodes. Guthrie discovered that a positively charged electroscope could be discharged by bringing a grounded piece of white-hot metal close to it (but not actually touching it). The same did not apply to a negatively charged electroscope, indicating that the current flow was only possible in one direction.

Thomas Edison independently rediscovered the principle on February 13, 1880. At the time, Edison was investigating why the filaments of his carbon-filament light bulbs nearly always burned out at the positive-connected end. He had a special bulb made with a metal plate sealed into the glass envelope. Using this device, he confirmed that an invisible current flowed from the glowing filament through the vacuum to the metal plate, but only when the plate was connected to the positive supply.

Edison devised a circuit where his modified light bulb effectively replaced the resistor in a DC voltmeter. Edison was awarded a patent for this invention in 1884. Since there was no apparent practical use for such a device at the time, the patent application was most likely simply a precaution in case someone else did find a use for the so-called Edison effect.

About 20 years later, John Ambrose Fleming (scientific adviser to the Marconi Company and former Edison employee) realized that the Edison effect could be used as a precision radio detector. Fleming patented the first true thermionic diode, the Fleming valve, in Britain on November 16, 1904 (followed by U.S. Patent 803,684 in November 1905).

Solid-state Diodes

In 1874 German scientist Karl Ferdinand Braun discovered the "unilateral conduction" of crystals. Braun patented the crystal rectifier in 1899. Copper oxide and selenium rectifiers were developed for power applications in the 1930s.

Indian scientist Jagadish Chandra Bose was the first to use a crystal for detecting radio waves in 1894. The crystal detector was developed into a practical device for wireless telegraphy by Greenleaf Whittier Pickard, who invented a silicon crystal detector in 1903 and received a patent for it on November 20, 1906. Other experimenters tried a variety of other substances, of which the most widely used was the mineral galena (lead sulfide). Other substances offered slightly better performance, but galena was most widely used because it had the advantage of being cheap and easy to obtain. The crystal detector in these early crystal radio sets consisted of an adjustable wire point-contact, often made of gold or platinum because of their incorrodible nature (the so-called "cat's whisker"), which could be manually moved over the face of the crystal in search of a portion of that mineral with rectifying qualties. This troublesome device was superseded by thermionic diodes (vacuum tubes) by the 1920s, but after high purity semiconductor materials became available, the crystal detector returned to dominant use with the advent, in the 1950s, of inexpensive fixed-germanium diodes. Bell Labs also developed a germanium diode for microwave reception, and AT&T used these in their microwave towers that criss-crossed the nation starting in the late 1940s, carrying telephone and network television signals. Bell Labs did not develop a satisfactory thermionic diode for microwave reception.

Etymology

At the time of their invention, such devices were known as rectifiers. In 1919, the year tetrodes were invented, William Henry Eccles coined the term *diode* from the Greek roots *di*, meaning 'two', and *ode*, meaning 'path'. (However, the word *diode* itself, as well as *triode, tetrode, pentode, hexode*, were already in use as terms of multiplex telegraphy).

Rectifiers

Although all diodes *rectify*, the term 'rectifier' is normally reserved for higher currents and voltages than would normally be found in the rectification of lower power signals; examples include:

- Power supply rectifiers (*half-wave, full-wave, bridge*)

- Flyback diodes

World's Smallest Diode

Researchers from the University of Georgia and Ben-Gurion University of the Negev (BGU) have developed a diode made from a molecule of DNA. Professor Bingqian Xu from the College of Engineering at the University of Georgia and his team took a single DNA molecule made from 11 base pairs and connected it to an electronic circuit a few nanometers in size. When layers of coralyne were inserted between layers of DNA, the current jumped up to 15 times larger negative versus positive, which is necessary for a nano diode.

Thermionic Diodes

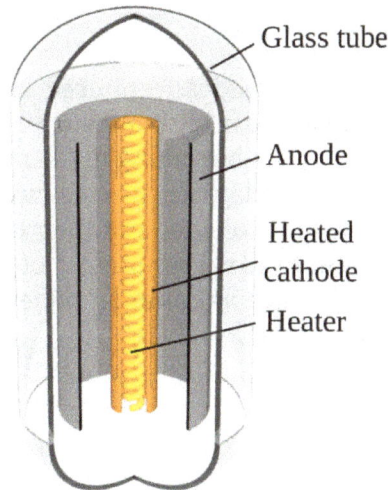

Diode vacuum tube construction

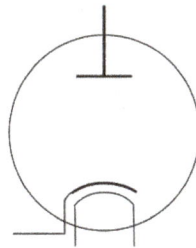

The symbol for an indirect heated vacuum-tube diode. From top to bottom, the components are the anode, the cathode, and the heater filament.

A thermionic diode is a thermionic-valve device (also known as a vacuum tube, tube, or valve), consisting of a sealed evacuated glass envelope containing two electrodes: a cathode heated by a filament, and a plate (anode). Early examples were fairly similar in appearance to incandescent light bulbs.

In operation, a separate current through the filament (heater), a high resistance wire made of nichrome, heats the cathode red hot (800–1000 °C), causing it to release electrons into the vacuum, a process called thermionic emission. The cathode is coated with oxides of alkaline earth metals such as barium and strontium oxides, which have a low work function, to increase the number of electrons emitted. (Some valves use *direct heating*, in which a tungsten filament acts as both heater and cathode.) The alternating voltage to be rectified is applied between the cathode and the concentric plate electrode. When the plate has a positive voltage with respect to the cathode, it electrostatically attracts the electrons from the cathode, so a current of electrons flows through the tube from cathode to plate. However, when the polarity is reversed and the plate has a negative voltage, no current flows, because the cathode electrons are not attracted to it. The unheated plate does not emit any electrons itself. So electrons can only flow through the tube in one direction, from the cathode to the anode plate.

In a mercury-arc valve, an arc forms between a refractory conductive anode and a pool of liquid mercury acting as cathode. Such units were made with ratings up to hundreds of kilowatts, and were important in the development of HVDC power transmission. Some types of smaller thermionic rectifiers had mercury vapor fill to reduce their forward voltage drop and to increase current rating over thermionic hard-vacuum devices.

Throughout the vacuum tube era, valve diodes were used in analog signal applications and as rectifiers in DC power supplies in consumer electronics such as radios, televisions, and sound systems. They were replaced in power supplies beginning in the 1940s by selenium rectifiers and then by semiconductor diodes by the 1960s. Today they are still used in a few high power applications where their ability to withstand transient voltages and their robustness gives them an advantage over semiconductor devices. The recent (2012) resurgence of interest among audiophiles and recording studios in old valve audio gear such as guitar amplifiers and home audio systems has provided a market for the legacy consumer diode valves.

Semiconductor Diodes

Electronic Symbols

The symbol used for a semiconductor diode in a circuit diagram specifies the type of diode. There are alternative symbols for some types of diodes, though the differences are minor. The triangle in the symbols points to the forward direction.

Anode   Cathode

Diode

Anode   Cathode

Light-emitting diode (LED)

Anode   Cathode

Photodiode

Anode   Cathode

Schottky diode

Transient-voltage-suppression diode (TVS)

Tunnel diode

Varicap

Zener diode

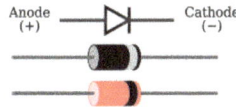

Typical diode packages in same alignment as diode symbol. Thin bar depicts the cathode.

A galena cat's-whisker detector, a point-contact diode.

Point-contact Diodes

A point-contact diode works the same as the junction diodes described below, but its construction is simpler. A pointed metal wire is placed in contact with an n-type semiconductor. Some metal migrates into the semiconductor to make a small p-type region around the contact. The 1N34 germanium version is still used in radio receivers as a detector and occasionally in specialized analog electronics.

Junction Diodes

p–n junction Diode

A p–n junction diode is made of a crystal of semiconductor, usually silicon, but germanium and gallium arsenide are also used. Impurities are added to it to create a region on one side that contains negative charge carriers (electrons), called an n-type semiconductor, and a region on the other side that contains positive charge carriers (holes), called a p-type semiconductor. When the n-type and p-type materials are attached together, a momentary flow of electrons occur from the n to the p side resulting in a third region between the two where no charge carriers are present. This region is called the depletion region because there are no charge carriers (neither electrons nor holes) in it. The diode's terminals are attached to the n-type and p-regions. The boundary between these two regions, called a p–n junction, is where the action of the diode takes place. When a sufficiently higher electrical potential is applied to the P side (the anode) than to the N side (the cathode), it allows electrons to flow through the depletion region from the N-type side to the P-type side. The junction does not allow the flow of electrons in the opposite direction when the potential is applied in reverse, creating, in a sense, an electrical check valve.

Schottky Diode

Another type of junction diode, the Schottky diode, is formed from a metal–semiconductor junction rather than a p–n junction, which reduces capacitance and increases switching speed.

Current–voltage Characteristic

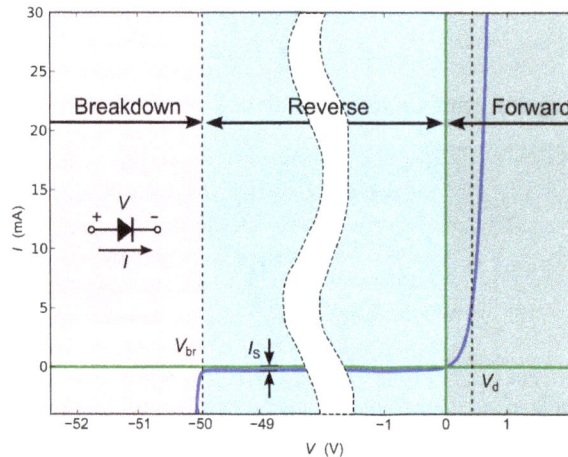

I–V (current vs. voltage) characteristics of a p–n junction diode

A semiconductor diode's behavior in a circuit is given by its current–voltage characteristic, or I–V graph. The shape of the curve is determined by the transport of charge carriers through the so-called *depletion layer* or *depletion region* that exists at the p–n junction between differing semiconductors. When a p–n junction is first created, conduction-band (mobile) electrons from the N-doped region diffuse into the P-doped region where there is a large population of holes (vacant places for electrons) with which the electrons "recombine". When a mobile electron recombines with a hole, both hole and electron vanish, leaving behind an immobile positively charged donor (dopant) on the N side and negatively charged acceptor (dopant) on the P side. The region around the p–n junction becomes depleted of charge carriers and thus behaves as an insulator.

However, the width of the depletion region (called the depletion width) cannot grow without limit. For each electron–hole pair recombination made, a positively charged dopant ion is left behind in the N-doped region, and a negatively charged dopant ion is created in the P-doped region. As recombination proceeds and more ions are created, an increasing electric field develops through the depletion zone that acts to slow and then finally stop recombination. At this point, there is a "built-in" potential across the depletion zone.

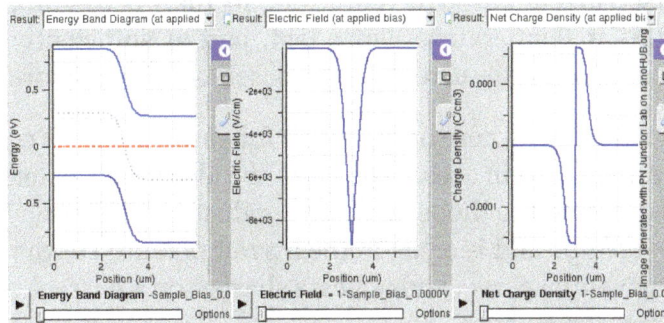

A PN junction diode in forward bias mode, the depletion width decreases. Both p and n junctions are doped at a 1e15/cm3 doping level, leading to built-in potential of ~0.59V. Observe the different Quasi Fermi levels for conduction band and valence band in n and p regions (red curves).

Reverse Bias

If an external voltage is placed across the diode with the same polarity as the built-in potential, the depletion zone continues to act as an insulator, preventing any significant electric current flow (unless electron–hole pairs are actively being created in the junction by, for instance, light). This is called the *reverse bias* phenomenon.

Forward Bias

However, if the polarity of the external voltage opposes the built-in potential, recombination can once again proceed, resulting in a substantial electric current through the p–n junction (i.e. substantial numbers of electrons and holes recombine at the junction). For silicon diodes, the built-in potential is approximately 0.7 V (0.3 V for germanium and 0.2 V for Schottky). Thus, if an external voltage greater than and opposite to the built-in voltage is applied, a current will flow and the diode is said to be "turned on" as it has been given an external *forward bias*. The diode is commonly said to have a forward "threshold" voltage, above which it conducts and below which conduction stops. However, this is only an approximation as the forward characteristic is according to the Shockley equation absolutely smooth.

A diode's I–V characteristic can be approximated by four regions of operation:

1. At very large reverse bias, beyond the peak inverse voltage or PIV, a process called reverse breakdown occurs that causes a large increase in current (i.e., a large number of electrons and holes are created at, and move away from the p–n junction) that usually damages the device permanently. The avalanche diode is deliberately designed for use in that manner. In the Zener diode, the concept of PIV is not applicable. A Zener diode contains a heavily doped p–n junction allowing electrons to tunnel from the valence band of the p-type material to the conduction band of the n-type material, such that the reverse voltage is "clamped" to a known value (called the *Zener voltage*), and avalanche does not occur. Both

devices, however, do have a limit to the maximum current and power they can withstand in the clamped reverse-voltage region. Also, following the end of forward conduction in any diode, there is reverse current for a short time. The device does not attain its full blocking capability until the reverse current ceases.

2. For a bias less than the PIV, the reverse current is very small. For a normal P–N rectifier diode, the reverse current through the device in the micro-ampere (µA) range is very low. However, this is temperature dependent, and at sufficiently high temperatures, a substantial amount of reverse current can be observed (mA or more).

3. With a small forward bias, where only a small forward current is conducted, the current–voltage curve is exponential in accordance with the ideal diode equation. There is a definite forward voltage at which the diode starts to conduct significantly. This is called the *knee voltage* or *cut-in voltage* and is equal to the barrier potential of the p-n junction. This is a feature of the exponential curve, and appears sharper on a current scale more compressed than in the diagram shown here.

4. At larger forward currents the current-voltage curve starts to be dominated by the ohmic resistance of the bulk semiconductor. The curve is no longer exponential, it is asymptotic to a straight line whose slope is the bulk resistance. This region is particularly important for power diodes. The diode can be modeled as an ideal diode in series with a fixed resistor.

In a small silicon diode operating at its rated currents, the voltage drop is about 0.6 to 0.7 volts. The value is different for other diode types—Schottky diodes can be rated as low as 0.2 V, germanium diodes 0.25 to 0.3 V, and red or blue light-emitting diodes (LEDs) can have values of 1.4 V and 4.0 V respectively.

At higher currents the forward voltage drop of the diode increases. A drop of 1 V to 1.5 V is typical at full rated current for power diodes.

Shockley Diode Equation

The *Shockley ideal diode equation* or the *diode law* (named after transistor co-inventor William Bradford Shockley) gives the I–V characteristic of an ideal diode in either forward or reverse bias (or no bias). The following equation is called the *Shockley ideal diode equation* when n, the ideality factor, is set equal to 1 :

$$I = I_S \left(e^{\frac{V_D}{nV_T}} - 1 \right)$$

where

I is the diode current,

I_S is the reverse bias saturation current (or scale current),

V_D is the voltage across the diode,

V_T is the thermal voltage, and

n is the *ideality factor*, also known as the *quality factor* or sometimes *emission coefficient*. The ideality factor *n* typically varies from 1 to 2 (though can in some cases be higher), depending on the fabrication process and semiconductor material and is set equal to 1 for the case of an "ideal" diode (thus the n is sometimes omitted). The ideality factor was added to account for imperfect junctions as observed in real transistors. The factor mainly accounts for carrier recombination as the charge carriers cross the depletion region.

The thermal voltage V_T is approximately 25.85 mV at 300 K, a temperature close to "room temperature" commonly used in device simulation software. At any temperature it is a known constant defined by:

$$V_T = \frac{kT}{q},$$

where *k* is the Boltzmann constant, *T* is the absolute temperature of the p–n junction, and *q* is the magnitude of charge of an electron (the elementary charge).

The reverse saturation current, I_S, is not constant for a given device, but varies with temperature; usually more significantly than V_T, so that V_D typically decreases as *T* increases.

The *Shockley ideal diode equation* or the *diode law* is derived with the assumption that the only processes giving rise to the current in the diode are drift (due to electrical field), diffusion, and thermal recombination–generation (R–G) (this equation is derived by setting n = 1 above). It also assumes that the R–G current in the depletion region is insignificant. This means that the *Shockley ideal diode equation* doesn't account for the processes involved in reverse breakdown and photon-assisted R–G. Additionally, it doesn't describe the "leveling off" of the I–V curve at high forward bias due to internal resistance. Introducing the ideality factor, n, accounts for recombination and generation of carriers.

Under *reverse bias* voltages the exponential in the diode equation is negligible, and the current is a constant (negative) reverse current value of $-I_S$. The reverse *breakdown region* is not modeled by the Shockley diode equation.

For even rather small *forward bias* voltages the exponential is very large, since the thermal voltage is very small in comparison. The subtracted '1' in the diode equation is then negligible and the forward diode current can be approximated by

$$I = I_S e^{\frac{V_D}{nV_T}}$$

The use of the diode equation in circuit problems is illustrated in the article on diode modeling.

Small-signal Behavior

For circuit design, a small-signal model of the diode behavior often proves useful. A specific example of diode modeling is discussed in the article on small-signal circuits.

Reverse-recovery Effect

Following the end of forward conduction in a p–n type diode, a reverse current can flow for a short time. The device does not attain its blocking capability until the mobile charge in the junction is depleted.

The effect can be significant when switching large currents very quickly. A certain amount of "reverse recovery time" t_r (on the order of tens of nanoseconds to a few microseconds) may be required to remove the reverse recovery charge Q_r from the diode. During this recovery time, the diode can actually conduct in the reverse direction. This might give rise to a large constant current in the reverse direction for a short time while the diode is reverse biased. The magnitude of such a reverse current is determined by the operating circuit (i.e., the series resistance) and the diode is said to be in the storage-phase. In certain real-world cases it is important to consider the losses that are incurred by this non-ideal diode effect. However, when the slew rate of the current is not so severe (e.g. Line frequency) the effect can be safely ignored. For most applications, the effect is also negligible for Schottky diodes.

The reverse current ceases abruptly when the stored charge is depleted; this abrupt stop is exploited in step recovery diodes for generation of extremely short pulses.

Types of Semiconductor Diode

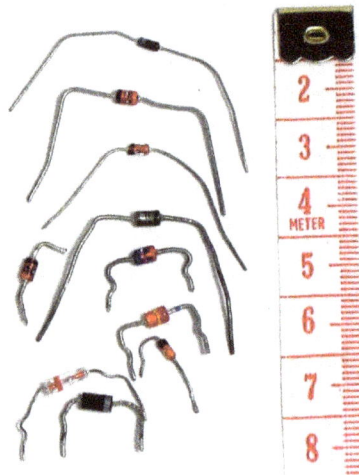

Several types of diodes. The scale is centimeters.

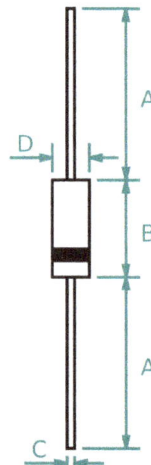

Typical datasheet drawing showing the dimensions of a DO-41 diode package

There are several types of p–n junction diodes, which emphasize either a different physical aspect of a diode often by geometric scaling, doping level, choosing the right electrodes, are just an application of a diode in a special circuit, or are really different devices like the Gunn and laser diode and the MOSFET:

Normal (p–n) diodes, which operate as described above, are usually made of doped silicon or, more rarely, germanium. Before the development of silicon power rectifier diodes, cuprous oxide and later selenium was used. Their low efficiency required a much higher forward voltage to be applied (typically 1.4 to 1.7 V per "cell", with multiple cells stacked so as to increase the peak inverse voltage rating for application in high voltage rectifiers), and required a large heat sink (often an extension of the diode's metal substrate), much larger than the later silicon diode of the same current ratings would require. The vast majority of all diodes are the p–n diodes found in CMOS integrated circuits, which include two diodes per pin and many other internal diodes.

Avalanche diodes

These are diodes that conduct in the reverse direction when the reverse bias voltage exceeds the breakdown voltage. These are electrically very similar to Zener diodes (and are often mistakenly called Zener diodes), but break down by a different mechanism: the *avalanche effect*. This occurs when the reverse electric field applied across the p–n junction causes a wave of ionization, reminiscent of an avalanche, leading to a large current. Avalanche diodes are designed to break down at a well-defined reverse voltage without being destroyed. The difference between the avalanche diode (which has a reverse breakdown above about 6.2 V) and the Zener is that the channel length of the former exceeds the mean free path of the electrons, resulting in many collisions between them on the way through the channel. The only practical difference between the two types is they have temperature coefficients of opposite polarities.

Cat's whisker or crystal diodes

These are a type of point-contact diode. The cat's whisker diode consists of a thin or sharpened metal wire pressed against a semiconducting crystal, typically galena or a piece of coal. The wire forms the anode and the crystal forms the cathode. Cat's whisker diodes were also called crystal diodes and found application in the earliest radios called crystal radio receivers. Cat's whisker diodes are generally obsolete, but may be available from a few manufacturers.

Constant current diodes

These are actually JFETs with the gate shorted to the source, and function like a two-terminal current-limiting analog to the voltage-limiting Zener diode. They allow a current through them to rise to a certain value, and then level off at a specific value. Also called *CLDs*, *constant-current diodes*, *diode-connected transistors*, or *current-regulating diodes*.

Esaki or tunnel diodes

These have a region of operation showing negative resistance caused by quantum tunneling, allowing amplification of signals and very simple bistable circuits. Because of the high carrier concentration, tunnel diodes are very fast, may be used at low (mK) temperatures, high magnetic fields, and in high radiation environments. Because of these properties, they are often used in spacecraft.

Gunn diodes

These are similar to tunnel diodes in that they are made of materials such as GaAs or InP that exhibit a region of negative differential resistance. With appropriate biasing, dipole domains form and travel across the diode, allowing high frequency microwave oscillators to be built.

Light-emitting diodes (LEDs)

In a diode formed from a direct band-gap semiconductor, such as gallium arsenide, charge carriers that cross the junction emit photons when they recombine with the majority carrier on the other side. Depending on the material, wavelengths (or colors) from the infrared to the near ultraviolet may be produced. The forward potential of these diodes depends on the wavelength of the emitted photons: 2.1 V corresponds to red, 4.0 V to violet. The first LEDs were red and yellow, and higher-frequency diodes have been developed over time. All LEDs produce incoherent, narrow-spectrum light; "white" LEDs are actually combinations of three LEDs of a different color, or a blue LED with a yellow scintillator coating. LEDs can also be used as low-efficiency photodiodes in signal applications. An LED may be paired with a photodiode or phototransistor in the same package, to form an opto-isolator.

Laser diodes

When an LED-like structure is contained in a resonant cavity formed by polishing the parallel end faces, a laser can be formed. Laser diodes are commonly used in optical storage devices and for high speed optical communication.

Thermal diodes

This term is used both for conventional p–n diodes used to monitor temperature because of their varying forward voltage with temperature, and for Peltier heat pumps for thermoelectric heating and cooling. Peltier heat pumps may be made from semiconductor, though they do not have any rectifying junctions, they use the differing behaviour of charge carriers in N and P type semiconductor to move heat.

Perun's diodes

This is a special type of voltage-surge protection diode. It is characterized by the symmetrical voltage-current characteristic, similar to DIAC. It has much faster response time however, that's why it is used in demanding applications.

Photodiodes

All semiconductors are subject to optical charge carrier generation. This is typically an undesired effect, so most semiconductors are packaged in light blocking material. Photodiodes are intended to sense light(photodetector), so they are packaged in materials that allow light to pass, and are usually PIN (the kind of diode most sensitive to light). A photodiode can be used in solar cells, in photometry, or in optical communications. Multiple photodiodes may be packaged in a single device, either as a linear array or as a two-dimensional array. These arrays should not be confused with charge-coupled devices.

PIN diodes

A PIN diode has a central un-doped, or *intrinsic*, layer, forming a p-type/intrinsic/n-type structure. They are used as radio frequency switches and attenuators. They are also used as large-volume, ionizing-radiation detectors and as photodetectors. PIN diodes are also used in power electronics, as their central layer can withstand high voltages. Furthermore, the PIN structure can be found in many power semiconductor devices, such as IGBTs, power MOSFETs, and thyristors.

Schottky diodes

Schottky diodes are constructed from a metal to semiconductor contact. They have a lower forward voltage drop than p–n junction diodes. Their forward voltage drop at forward currents of about 1 mA is in the range 0.15 V to 0.45 V, which makes them useful in voltage clamping applications and prevention of transistor saturation. They can also be used as low loss rectifiers, although their reverse leakage current is in general higher than that of other diodes. Schottky diodes are majority carrier devices and so do not suffer from minority carrier storage problems that slow down many other diodes—so they have a faster reverse recovery than p–n junction diodes. They also tend to have much lower junction capacitance than p–n diodes, which provides for high switching speeds and their use in high-speed circuitry and RF devices such as switched-mode power supply, mixers, and detectors.

Super barrier diodes

Super barrier diodes are rectifier diodes that incorporate the low forward voltage drop of the Schottky diode with the surge-handling capability and low reverse leakage current of a normal p–n junction diode.

Gold-doped diodes

As a dopant, gold (or platinum) acts as recombination centers, which helps a fast recombination of minority carriers. This allows the diode to operate at signal frequencies, at the expense of a higher forward voltage drop. Gold-doped diodes are faster than other p–n diodes (but not as fast as Schottky diodes). They also have less reverse-current leakage than Schottky diodes (but not as good as other p–n diodes). A typical example is the 1N914.

Snap-off or Step recovery diodes

The term *step recovery* relates to the form of the reverse recovery characteristic of these devices. After a forward current has been passing in an SRD and the current is interrupted or reversed, the reverse conduction will cease very abruptly (as in a step waveform). SRDs can, therefore, provide very fast voltage transitions by the very sudden disappearance of the charge carriers.

Stabistors or *Forward Reference Diodes*

The term *stabistor* refers to a special type of diodes featuring extremely stable forward voltage characteristics. These devices are specially designed for low-voltage stabilization applications requiring a guaranteed voltage over a wide current range and highly stable over temperature.

Transient voltage suppression diode (TVS)

> These are avalanche diodes designed specifically to protect other semiconductor devices from high-voltage transients. Their p–n junctions have a much larger cross-sectional area than those of a normal diode, allowing them to conduct large currents to ground without sustaining damage.

Varicap or varactor diodes

> These are used as voltage-controlled capacitors. These are important in PLL (phase-locked loop) and FLL (frequency-locked loop) circuits, allowing tuning circuits, such as those in television receivers, to lock quickly on to the frequency. They also enabled tunable oscillators in early discrete tuning of radios, where a cheap and stable, but fixed-frequency, crystal oscillator provided the reference frequency for a voltage-controlled oscillator.

Zener diodes

> These can be made to conduct in reverse bias (backward), and are correctly termed reverse breakdown diodes. This effect, called Zener breakdown, occurs at a precisely defined voltage, allowing the diode to be used as a precision voltage reference. The term Zener diode is colloquially applied to several types of breakdown diodes, but strictly speaking Zener diodes have a breakdown voltage of below 5 volts, whilst avalanche diodes are used for breakdown voltages above that value. In practical voltage reference circuits, Zener and switching diodes are connected in series and opposite directions to balance the temperature coefficient response of the diodes to near-zero. Some devices labeled as high-voltage Zener diodes are actually avalanche diodes. Two (equivalent) Zeners in series and in reverse order, in the same package, constitute a transient absorber (or Transorb, a registered trademark).

Other uses for semiconductor diodes include the sensing of temperature, and computing analog logarithms.

Numbering and Coding Schemes

There are a number of common, standard and manufacturer-driven numbering and coding schemes for diodes; the two most common being the EIA/JEDEC standard and the European Pro Electron standard:

EIA/JEDEC

The standardized 1N-series numbering *EIA370* system was introduced in the US by EIA/JEDEC (Joint Electron Device Engineering Council) about 1960. Most diodes have a 1-prefix designation (e.g., 1N4003). Among the most popular in this series were: 1N34A/1N270 (germanium signal), 1N914/1N4148 (silicon signal), 1N400x (silicon 1A power rectifier), and 1N580x (silicon 3A power rectifier).

JIS

The JIS semiconductor designation system has all semiconductor diode designations starting with "1S".

Pro Electron

The European Pro Electron coding system for active components was introduced in 1966 and comprises two letters followed by the part code. The first letter represents the semiconductor material used for the component (A = germanium and B = silicon) and the second letter represents the general function of the part (for diodes, A = low-power/signal, B = variable capacitance, X = multiplier, Y = rectifier and Z = voltage reference); for example:

- AA-series germanium low-power/signal diodes (e.g., AA119)

- BA-series silicon low-power/signal diodes (e.g., BAT18 silicon RF switching diode)

- BY-series silicon rectifier diodes (e.g., BY127 1250V, 1A rectifier diode)

- BZ-series silicon Zener diodes (e.g., BZY88C4V7 4.7V Zener diode)

Other common numbering / coding systems (generally manufacturer-driven) include:

- GD-series germanium diodes (e.g., GD9) – this is a very old coding system

- OA-series germanium diodes (e.g., OA47) – a coding sequence developed by Mullard, a UK company

As well as these common codes, many manufacturers or organisations have their own systems too – for example:

- HP diode 1901-0044 = JEDEC 1N4148

- UK military diode CV448 = Mullard type OA81 = GEC type GEX23

Related devices

- Rectifier

- Transistor

- Thyristor or silicon controlled rectifier (SCR)

- TRIAC

- DIAC

- Varistor

In optics, an equivalent device for the diode but with laser light would be the Optical isolator, also known as an Optical Diode, that allows light to only pass in one direction. It uses a Faraday rotator as the main component.

Applications

Radio Demodulation

The first use for the diode was the demodulation of amplitude modulated (AM) radio broadcasts. The history of this discovery is treated in depth in the radio article. In summary, an AM signal consists of alternating positive and negative peaks of a radio carrier wave, whose amplitude or envelope is proportional to the original audio signal. The diode (originally a crystal diode) rectifies the AM radio frequency signal, leaving only the positive peaks of the carrier wave. The audio is then extracted from the rectified carrier wave using a simple filter and fed into an audio amplifier or transducer, which generates sound waves.

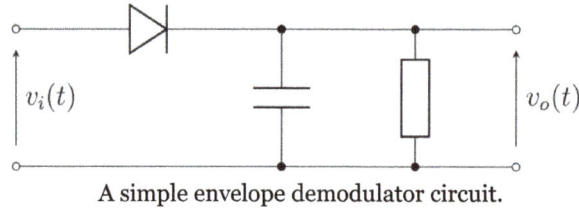

A simple envelope demodulator circuit.

Power Conversion

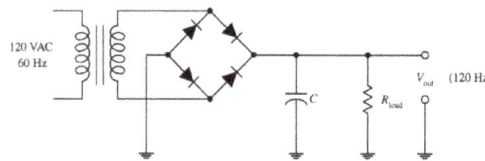

Schematic of basic AC-to-DC power supply

Rectifiers are constructed from diodes, where they are used to convert alternating current (AC) electricity into direct current (DC). Automotive alternators are a common example, where the diode, which rectifies the AC into DC, provides better performance than the commutator or earlier, dynamo. Similarly, diodes are also used in *Cockcroft–Walton voltage multipliers* to convert AC into higher DC voltages.

Over-voltage Protection

Diodes are frequently used to conduct damaging high voltages away from sensitive electronic devices. They are usually reverse-biased (non-conducting) under normal circumstances. When the voltage rises above the normal range, the diodes become forward-biased (conducting). For example, diodes are used in (stepper motor and H-bridge) motor controller and relay circuits to de-energize coils rapidly without the damaging voltage spikes that would otherwise occur. (A diode used in such an application is called a flyback diode). Many integrated circuits also incorporate diodes on the connection pins to prevent external voltages from damaging their sensitive transistors. Specialized diodes are used to protect from over-voltages at higher power.

Logic Gates

Diodes can be combined with other components to construct AND and OR logic gates. This is referred to as diode logic.

Ionizing Radiation Detectors

In addition to light, mentioned above, semiconductor diodes are sensitive to more energetic radiation. In electronics, cosmic rays and other sources of ionizing radiation cause noise pulses and single and multiple bit errors. This effect is sometimes exploited by particle detectors to detect radiation. A single particle of radiation, with thousands or millions of electron volts of energy, generates many charge carrier pairs, as its energy is deposited in the semiconductor material. If the depletion layer is large enough to catch the whole shower or to stop a heavy particle, a fairly accurate measurement of the particle's energy can be made, simply by measuring the charge conducted and without the complexity of a magnetic spectrometer, etc. These semiconductor radiation detectors need efficient and uniform charge collection and low leakage current. They are often cooled by liquid nitrogen. For longer-range (about a centimetre) particles, they need a very large depletion depth and large area. For short-range particles, they need any contact or un-depleted semiconductor on at least one surface to be very thin. The back-bias voltages are near breakdown (around a thousand volts per centimetre). Germanium and silicon are common materials. Some of these detectors sense position as well as energy. They have a finite life, especially when detecting heavy particles, because of radiation damage. Silicon and germanium are quite different in their ability to convert gamma rays to electron showers.

Semiconductor detectors for high-energy particles are used in large numbers. Because of energy loss fluctuations, accurate measurement of the energy deposited is of less use.

Temperature Measurements

A diode can be used as a temperature measuring device, since the forward voltage drop across the diode depends on temperature, as in a silicon bandgap temperature sensor. From the Shockley ideal diode equation given above, it might *appear* that the voltage has a *positive* temperature coefficient (at a constant current), but usually the variation of the reverse saturation current term is more significant than the variation in the thermal voltage term. Most diodes therefore have a *negative* temperature coefficient, typically -2 mV/$^\circ$C for silicon diodes. The temperature coefficient is approximately constant for temperatures above about 20 kelvins. Some graphs are given for 1N400x series, and CY7 cryogenic temperature sensor.

Current Steering

Diodes will prevent currents in unintended directions. To supply power to an electrical circuit during a power failure, the circuit can draw current from a battery. An uninterruptible power supply may use diodes in this way to ensure that current is only drawn from the battery when necessary. Likewise, small boats typically have two circuits each with their own battery/batteries: one used for engine starting; one used for domestics. Normally, both are charged from a single alternator, and a heavy-duty split-charge diode is used to prevent the higher-charge battery (typically the engine battery) from discharging through the lower-charge battery when the alternator is not running.

Diodes are also used in electronic musical keyboards. To reduce the amount of wiring needed in electronic musical keyboards, these instruments often use keyboard matrix circuits. The keyboard controller scans the rows and columns to determine which note the player has pressed. The problem with matrix circuits is that, when several notes are pressed at once, the current can flow backwards through the circuit and trigger "phantom keys" that cause "ghost" notes to play. To avoid triggering unwanted notes, most keyboard matrix circuits have diodes soldered with the switch under each key of the musical keyboard. The same principle is also used for the switch matrix in solid-state pinball machines.

Waveform Clipper

Diodes can be used to limit the positive or negative excursion of a signal to a prescribed voltage.

Clamper

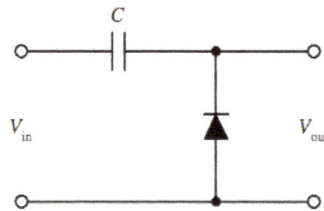

This simple diode clamp will clamp the negative peaks of the incoming waveform to the common rail voltage

A diode clamp circuit can take a periodic alternating current signal that oscillates between positive and negative values, and vertically displace it such that either the positive, or the negative peaks occur at a prescribed level. The clamper does not restrict the peak-to-peak excursion of the signal, it moves the whole signal up or down so as to place the peaks at the reference level.

Abbreviations

Diodes are usually referred to as *D* for diode on PCBs. Sometimes the abbreviation *CR* for *crystal rectifier* is used.

References

- Pugh, Emerson W.; Johnson, Lyle R.; Palmer, John H. (1991). IBM's 360 and early 370 systems. MIT Press. p. 34. ISBN 0-262-16123-0.

- Paul Horowitz and Winfield Hill, The Art of Electronics 2nd Ed. Cambridge University Press, Cambridge, 1989 ISBN 0-521-37095-7 page 471

- Singmin, Andrew (2001). Beginning Digital Electronics Through Projects. Newnes. p. 9. ISBN 0-7506-7269-2. Signals come from transducers...

- Miller, Mark R. (2002). Electronics the Easy Way. Barron's Educational Series. pp. 232–239. ISBN 0-7641-1981-8. Until the radio came along...

- Carr, Joseph J. (2000). Secrets of RF circuit design. McGraw-Hill Professional. p. 423. ISBN 0-07-137067-6. It is common in microwave systems...

- Chen, Wai-Kai (2005). The electrical engineering handbook. Academic Press. p. 101. ISBN 0-12-170960-4. Noise from an analog (or small-signal) perspective...

- Scherz, Paul (2006). Practical electronics for inventors. McGraw-Hill Professional. p. 730. ISBN 0-07-145281-8. In order for analog devices... to communicate with digital circuits...

- Williams, Jim (1991). Analog circuit design. Newnes. p. 238. ISBN 0-7506-9640-0. Even within companies producing both analog and digital products...

- Muller, Richard S. & Theodore I. Kamins (1986). Device Electronics for Integrated Circuits. John Wiley and Sons. ISBN 0-471-88758-7.

- Morris, C. G. (ed) (1992) Academic Press Dictionary of Science and Technology. Gulf Professional Publishing. p. 360. ISBN 0122004000.

- Jung, Walt. "Chapter 7 – Hardware and Housekeeping Techniques" (PDF). Op Amp Applications Handbook. p. 7.11. ISBN 0-7506-7844-5.

- Chelikowski, J. (2004) "Introduction: Silicon in all its Forms", p. 1 in Silicon: evolution and future of a technology. P. Siffert and E. F. Krimmel (eds.). Springer, ISBN 3-540-40546-1.

- McFarland, Grant (2006) Microprocessor design: a practical guide from design planning to manufacturing. McGraw-Hill Professional. p. 10. ISBN 0-07-145951-0.

- Heywang, W. and Zaininger, K. H. (2004) "Silicon: The Semiconductor Material", p. 36 in Silicon: evolution and future of a technology. P. Siffert and E. F. Krimmel (eds.). Springer, 2004 ISBN 3-540-40546-1.

- Price, Robert W. (2004). Roadmap to Entrepreneurial Success. AMACOM Div American Mgmt Assn. p. 42. ISBN 978-0-8144-7190-6.

- Kaplan, Daniel (2003). Hands-On Electronics. New York: Cambridge University Press. pp. 47–54, 60–61. ISBN 978-0-511-07668-8.

- Zhong Yuan Chang, Willy M. C. Sansen, Low-Noise Wide-Band Amplifiers in Bipolar and CMOS Technologies, page 31, Springer, 1991 ISBN 0792390962.

- Sedra, A.S. & Smith, K.C. (2004). Microelectronic circuits (Fifth ed.). New York: Oxford University Press. p. 397 and Figure 5.17. ISBN 0-19-514251-9.

- Bird, John (2010). Electrical and Electronic Principles and Technology. Routledge. pp. 63–76. ISBN 9780080890562. Retrieved 2013-03-17.

- Keithley, Joseph F. (1999). The Story of Electrical and Magnetic Measurements: From 500 BC to the 1940s. John Wiley & Sons. p. 23. ISBN 9780780311930. Retrieved 2013-03-17.

- Isaacson, Walter (2003). Benjamin Franklin: An American Life. Simon and Schuster. p. 136. ISBN 9780743260848. Retrieved 2013-03-17.

- Pai, S. T.; Qi Zhang (1995). Introduction to High Power Pulse Technology. Advanced Series in Electrical and Computer Engineering. 10. World Scientific. ISBN 9789810217143. Retrieved 2013-03-17.

- Dyer, Stephen A. (2004). Wiley Survey of Instrumentation and Measurement. John Wiley & Sons. p. 397. ISBN 9780471221654. Retrieved 2013-03-17.

- Scherz, Paul (2006). Practical Electronics for Inventors (2nd ed.). McGraw Hill Professional. p. 100. ISBN 9780071776448. Retrieved 2013-03-17.

Allied Fields of Electronic Circuit

The field of engineering that studies the functions of electricity, electronics and electromagnetism is known as electrical engineering. Electronic engineering particularly concentrates on devices, microprocessors and microcontrollers. The following text helps the readers to understand all the allied fields of electronic circuit.

Electrical Engineering

Electrical engineering is a field of engineering that generally deals with the study and application of electricity, electronics, and electromagnetism. This field first became an identifiable occupation in the later half of the 19th century after commercialization of the electric telegraph, the telephone, and electric power distribution and use. Subsequently, broadcasting and recording media made electronics part of daily life. The invention of the transistor, and later the integrated circuit, brought down the cost of electronics to the point they can be used in almost any household object.

Electrical engineers design complex power systems on a macroscopic level as well as microscopic electronic devices

Electrical engineering has now subdivided into a wide range of subfields including electronics, digital computers, power engineering, telecommunications, control systems, radio-frequency engineering, signal processing, instrumentation, and microelectronics. The subject of electronic engineering is often treated as its own subfield but it intersects with all the other subfields, including the power electronics of power engineering.

Electrical engineers typically hold a degree in electrical engineering or electronic engineering. Practicing engineers may have professional certification and be members of a professional body. Such bodies include the Institute of Electrical and Electronics Engineers (IEEE) and the Institution of Engineering and Technology (professional society) (IET).

Electrical engineers work in a very wide range of industries and the skills required are likewise variable. These range from basic circuit theory to the management skills required of a project manager. The tools and equipment that an individual engineer may need are similarly variable, ranging from a simple voltmeter to a top end analyzer to sophisticated design and manufacturing software.

History

Electricity has been a subject of scientific interest since at least the early 17th century. A prominent early electrical scientist was William Gilbert who was the first to draw a clear distinction between magnetism and static electricity and is credited with establishing the term electricity. He also designed the versorium: a device that detected the presence of statically charged objects. Then in 1762 Swedish professor Johan Carl Wilcke invented, and in 1775 Alessandro Volta improved, a device (for which Volta coined the name electrophorus) that produced a static electric charge, and by 1800 Volta had developed the voltaic pile, a forerunner of the electric battery.

19th Century

The discoveries of Michael Faraday formed the foundation of electric motor technology

In the 19th century, research into the subject started to intensify. Notable developments in this century include the work of Georg Ohm, who in 1827 quantified the relationship between the electric current and potential difference in a conductor, of Michael Faraday, the discoverer of electromagnetic induction in 1831, and of James Clerk Maxwell, who in 1873 published a unified theory of electricity and magnetism in his treatise *Electricity and Magnetism*.

Electrical engineering became a profession in the later 19th century. Practitioners had created a global electric telegraph network and the first professional electrical engineering institutions were founded in the UK and USA to support the new discipline. Although it is impossible to precisely pinpoint a first electrical engineer, Francis Ronalds stands ahead of the field, who created the first working electric telegraph system in 1816 and documented his vision of how the world could be transformed by electricity. Over 50 years later, he joined the new Society of Telegraph Engineers (soon to be renamed the Institution of Electrical Engineers) where he was regarded by other members as the first of their cohort. By the end of the 19th century, the world had been forever changed by the rapid communication made possible by the engineering development of land-lines, submarine cables, and, from about 1890, wireless telegraphy.

Practical applications and advances in such fields created an increasing need for standardised units of measure. They led to the international standardization of the units volt, ampere, coulomb, ohm, farad, and henry. This was achieved at an international conference in Chicago in 1893. The publication of these standards formed the basis of future advances in standardisation in various industries, and in many countries the definitions were immediately recognised in relevant legislation.

During these years, the study of electricity was largely considered to be a subfield of physics. That's because early electrical technology was electromechanical in nature. The Technische Universität Darmstadt founded the world's first department of electrical engineering in 1882. The first electrical engineering degree program was started at Massachusetts Institute of Technology (MIT) in the physics department under Professor Charles Cross, though it was Cornell University to produce the world's first electrical engineering graduates in 1885. The first course in electrical engineering was taught in 1883 in Cornell's Sibley College of Mechanical Engineering and Mechanic Arts. It was not until about 1885 that Cornell President Andrew Dickson White established the first Department of Electrical Engineering in the United States. In the same year, University College London founded the first chair of electrical engineering in Great Britain. Professor Mendell P. Weinbach at University of Missouri soon followed suit by establishing the electrical engineering department in 1886. Afterwards, universities and institutes of technology gradually started to offer electrical engineering programs to their students all over the world.

Thomas Edison, electric light and (DC) power supply networks

Károly Zipernowsky, Ottó Bláthy, Miksa Déri, the ZDB transformer

William Stanley, Jr., transformers

Galileo Ferraris, electrical theory, induction motor

Nikola Tesla, practical polyphase (AC) and induction motor designs

Mikhail Dolivo-Dobrovolsky developed standard 3-phase (AC) systems

Charles Proteus Steinmetz, AC mathematical theories for engineers

Oliver Heaviside, developed theoretical models for electric circuits

During these decades use of electrical engineering increased dramatically. In 1882, Thomas Edison switched on the world's first large-scale electric power network that provided 110 volts — direct current (DC) — to 59 customers on Manhattan Island in New York City. In 1884, Sir

Charles Parsons invented the steam turbine allowing for more efficient electric power generation. Alternating current, with its ability to transmit power more efficiently over long distances via the use of transformers, developed rapidly in the 1880s and 1890s with transformer designs by Károly Zipernowsky, Ottó Bláthy and Miksa Déri (later called ZBD transformers), Lucien Gaulard, John Dixon Gibbs and William Stanley, Jr.. Practical AC motor designs including induction motors were independently invented by Galileo Ferraris and Nikola Tesla and further developed into a practical three-phase form by Mikhail Dolivo-Dobrovolsky and Charles Eugene Lancelot Brown. Charles Steinmetz and Oliver Heaviside contributed to the theoretical basis of alternating current engineering. The spread in the use of AC set off in the United States what has been called the *War of Currents* between a George Westinghouse backed AC system and a Thomas Edison backed DC power system, with AC being adopted as the overall standard.

More Modern Developments

Guglielmo Marconi known for his pioneering work on long distance radio transmission

During the development of radio, many scientists and inventors contributed to radio technology and electronics. The mathematical work of James Clerk Maxwell during the 1850s had shown the relationship of different forms of electromagnetic radiation including possibility of invisible airborne waves (later called "radio waves"). In his classic physics experiments of 1888, Heinrich Hertz proved Maxwell's theory by transmitting radio waves with a spark-gap transmitter, and detected them by using simple electrical devices. Other physicists experimented with these new waves and in the process developed devices for transmitting and detecting them. In 1895, Guglielmo Marconi began work on a way to adapt the known methods of transmitting and detecting these "Hertzian waves" into a purpose built commercial wireless telegraphic system. Early on, he sent wireless signals over a distance of one and a half miles. In December 1901, he sent wireless waves that were not affected by the curvature of the Earth. Marconi later transmitted the wireless signals across the Atlantic between Poldhu, Cornwall, and St. John's, Newfoundland, a distance of 2,100 miles (3,400 km).

In 1897, Karl Ferdinand Braun introduced the cathode ray tube as part of an oscilloscope, a crucial enabling technology for electronic television. John Fleming invented the first radio tube, the diode, in 1904. Two years later, Robert von Lieben and Lee De Forest independently developed the amplifier tube, called the triode.

In 1920, Albert Hull developed the magnetron which would eventually lead to the development of the microwave oven in 1946 by Percy Spencer. In 1934, the British military began to make strides toward radar (which also uses the magnetron) under the direction of Dr Wimperis, culminating in the operation of the first radar station at Bawdsey in August 1936.

In 1941, Konrad Zuse presented the Z3, the world's first fully functional and programmable computer using electromechanical parts. In 1943, Tommy Flowers designed and built the Colossus, the world's first fully functional, electronic, digital and programmable computer. In 1946, the ENIAC (Electronic Numerical Integrator and Computer) of John Presper Eckert and John Mauchly followed, beginning the computing era. The arithmetic performance of these machines allowed engineers to develop completely new technologies and achieve new objectives, including the Apollo program which culminated in landing astronauts on the Moon.

Solid-state Transistors

The invention of the transistor in late 1947 by William B. Shockley, John Bardeen, and Walter Brattain of the Bell Telephone Laboratories opened the door for more compact devices and led to the development of the integrated circuit in 1958 by Jack Kilby and independently in 1959 by Robert Noyce. Starting in 1968, Ted Hoff and a team at the Intel Corporation invented the first commercial microprocessor, which foreshadowed the personal computer. The Intel 4004 was a four-bit processor released in 1971, but in 1973 the Intel 8080, an eight-bit processor, made the first personal computer, the Altair 8800, possible.

Subdisciplines

Electrical engineering has many subdisciplines, the most common of which are listed below. Although there are electrical engineers who focus exclusively on one of these subdisciplines, many deal with a combination of them. Sometimes certain fields, such as electronic engineering and computer engineering, are considered separate disciplines in their own right.

Power

Power pole

Power engineering deals with the generation, transmission, and distribution of electricity as well as the design of a range of related devices. These include transformers, electric generators, electric motors, high voltage engineering, and power electronics. In many regions of the world, governments maintain an electrical network called a power grid that connects a variety of generators together with users of their energy. Users purchase electrical energy from the grid, avoiding the costly exercise of having to generate their own. Power engineers may work on the design and maintenance of the power grid as well as the power systems that connect to it. Such systems are called *on-grid* power systems and may supply the grid with additional power, draw power from the grid, or do both. Power engineers may also work on systems that do not connect to the grid, called *off-grid* power systems, which in some cases are preferable to on-grid systems. The future includes Satellite controlled power systems, with feedback in real time to prevent power surges and prevent blackouts.

Control

Control systems play a critical role in spaceflight.

Control engineering focuses on the modeling of a diverse range of dynamic systems and the design of controllers that will cause these systems to behave in the desired manner. To implement such controllers, electrical engineers may use electronic circuits, digital signal processors, microcontrollers, and programmable logic controls (PLCs). Control engineering has a wide range of applications from the flight and propulsion systems of commercial airliners to the cruise control present in many modern automobiles. It also plays an important role in industrial automation.

Control engineers often utilize feedback when designing control systems. For example, in an automobile with cruise control the vehicle's speed is continuously monitored and fed back to the system which adjusts the motor's power output accordingly. Where there is regular feedback, control theory can be used to determine how the system responds to such feedback.

Electronics

Electronic engineering involves the design and testing of electronic circuits that use the properties of components such as resistors, capacitors, inductors, diodes, and transistors to achieve a particular functionality. The tuned circuit, which allows the user of a radio to filter out all but a single station, is just one example of such a circuit. Another example (of a pneumatic signal conditioner) is shown in the adjacent photograph.

Prior to the Second World War, the subject was commonly known as *radio engineering* and basically was restricted to aspects of communications and radar, commercial radio, and early television. Later, in post war years, as consumer devices began to be developed, the field grew to include modern television, audio systems, computers, and microprocessors. In the mid-to-late 1950s, the term *radio engineering* gradually gave way to the name *electronic engineering*.

Before the invention of the integrated circuit in 1959, electronic circuits were constructed from discrete components that could be manipulated by humans. These discrete circuits consumed much space and power and were limited in speed, although they are still common in some applications. By contrast, integrated circuits packed a large number—often millions—of tiny electrical components, mainly transistors, into a small chip around the size of a coin. This allowed for the powerful computers and other electronic devices we see today.

Microelectronics

Microelectronics engineering deals with the design and microfabrication of very small electronic circuit components for use in an integrated circuit or sometimes for use on their own as a general electronic component. The most common microelectronic components are semiconductor transistors, although all main electronic components (resistors, capacitors etc.) can be created at a microscopic level. Nanoelectronics is the further scaling of devices down to nanometer levels. Modern devices are already in the nanometer regime, with below 100 nm processing having been standard since about 2002.

Microelectronic components are created by chemically fabricating wafers of semiconductors such as silicon (at higher frequencies, compound semiconductors like gallium arsenide and indium phosphide) to obtain the desired transport of electronic charge and control of current. The field of microelectronics involves a significant amount of chemistry and material science and requires the electronic engineer working in the field to have a very good working knowledge of the effects of quantum mechanics.

Signal Processing

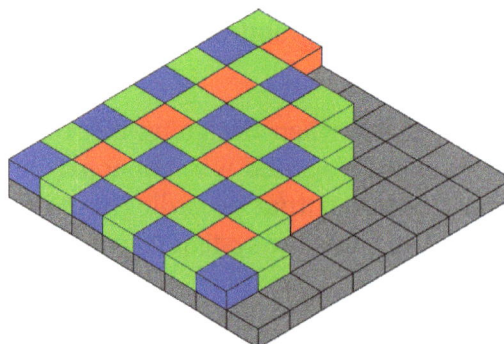

A Bayer filter on a CCD requires signal processing to get a red, green, and blue value at each pixel.

Signal processing deals with the analysis and manipulation of signals. Signals can be either analog, in which case the signal varies continuously according to the information, or digital, in which case the signal varies according to a series of discrete values representing the information. For analog signals, signal processing may involve the amplification and filtering of audio signals for audio equipment or the modulation and demodulation of signals for telecommunications. For digital signals, signal processing may involve the compression, error detection and error correction of digitally sampled signals.

Signal Processing is a very mathematically oriented and intensive area forming the core of digital signal processing and it is rapidly expanding with new applications in every field of electrical engineering such as communications, control, radar, audio engineering, broadcast engineering, power electronics, and biomedical engineering as many already existing analog systems are replaced with their digital counterparts. Analog signal processing is still important in the design of many control systems.

DSP processor ICs are found in every type of modern electronic systems and products including, SDTV | HDTV sets, radios and mobile communication devices, Hi-Fi audio equipment, Dolby noise reduction algorithms, GSM mobile phones, mp3 multimedia players, camcorders and digital cameras, automobile control systems, noise cancelling headphones, digital spectrum analyzers, intelligent missile guidance, radar, GPS based cruise control systems, and all kinds of image processing, video processing, audio processing, and speech processing systems.

Telecommunications

Satellite dishes are a crucial component in the analysis of satellite information.

Telecommunications engineering focuses on the transmission of information across a channel such as a coax cable, optical fiber or free space. Transmissions across free space require information to be encoded in a carrier signal to shift the information to a carrier frequency suitable for transmission; this is known as modulation. Popular analog modulation techniques include amplitude modulation and frequency modulation. The choice of modulation affects the cost and performance of a system and these two factors must be balanced carefully by the engineer.

Once the transmission characteristics of a system are determined, telecommunication engineers design the transmitters and receivers needed for such systems. These two are sometimes combined to form a two-way communication device known as a transceiver. A key consideration in the design of transmitters is their power consumption as this is closely related to their signal strength. If the signal strength of a transmitter is insufficient the signal's information will be corrupted by noise.

Instrumentation

Flight instruments provide pilots with the tools to control aircraft analytically.

Instrumentation engineering deals with the design of devices to measure physical quantities such as pressure, flow, and temperature. The design of such instrumentation requires a good understanding of physics that often extends beyond electromagnetic theory. For example, flight instruments measure variables such as wind speed and altitude to enable pilots the control of aircraft analytically. Similarly, thermocouples use the Peltier-Seebeck effect to measure the temperature difference between two points.

Often instrumentation is not used by itself, but instead as the sensors of larger electrical systems. For example, a thermocouple might be used to help ensure a furnace's temperature remains constant. For this reason, instrumentation engineering is often viewed as the counterpart of control engineering.

Computers

Supercomputers are used in fields as diverse as computational biology and geographic information systems.

Computer engineering deals with the design of computers and computer systems. This may involve the design of new hardware, the design of PDAs, tablets, and supercomputers, or the use of computers to control an industrial plant. Computer engineers may also work on a system's software. However, the design of complex software systems is often the domain of software engineering, which is usually considered a separate discipline. Desktop computers represent a tiny fraction of the devices a computer engineer might work on, as computer-like architectures are now found in a range of devices including video game consoles and DVD players.

Related Disciplines

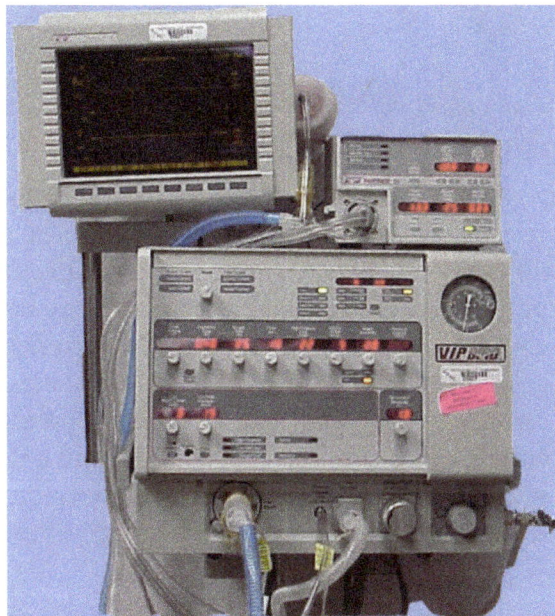

The Bird VIP Infant ventilator

Mechatronics is an engineering discipline which deals with the convergence of electrical and mechanical systems. Such combined systems are known as electromechanical systems and have widespread adoption. Examples include automated manufacturing systems, heating, ventilation and air-conditioning systems, and various subsystems of aircraft and automobiles.

The term *mechatronics* is typically used to refer to macroscopic systems but futurists have predicted the emergence of very small electromechanical devices. Already, such small devices, known as Microelectromechanical systems (MEMS), are used in automobiles to tell airbags when to deploy, in digital projectors to create sharper images, and in inkjet printers to create nozzles for high definition printing. In the future it is hoped the devices will help build tiny implantable medical devices and improve optical communication.

Biomedical engineering is another related discipline, concerned with the design of medical equipment. This includes fixed equipment such as ventilators, MRI scanners, and electrocardiograph monitors as well as mobile equipment such as cochlear implants, artificial pacemakers, and artificial hearts.

Aerospace engineering and robotics an example is the most recent electric propulsion and ion propulsion.

Education

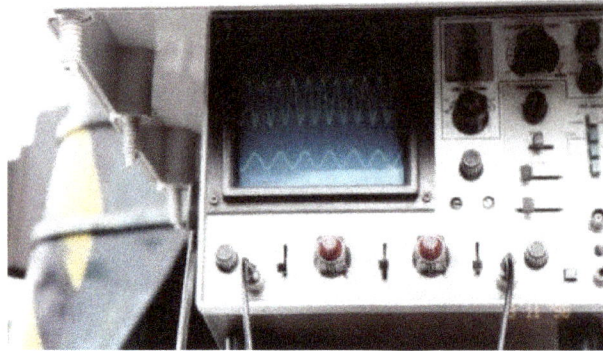

Oscilloscope

Electrical engineers typically possess an academic degree with a major in electrical engineering, electronics engineering, electrical engineering technology, or electrical and electronic engineering. The same fundamental principles are taught in all programs, though emphasis may vary according to title. The length of study for such a degree is usually four or five years and the completed degree may be designated as a Bachelor of Science in Electrical/Electronics Engineering Technology, Bachelor of Engineering, Bachelor of Science, Bachelor of Technology, or Bachelor of Applied Science depending on the university. The bachelor's degree generally includes units covering physics, mathematics, computer science, project management, and a variety of topics in electrical engineering. Initially such topics cover most, if not all, of the subdisciplines of electrical engineering. At some schools, the students can then choose to emphasize one or more subdisciplines towards the end of their courses of study.

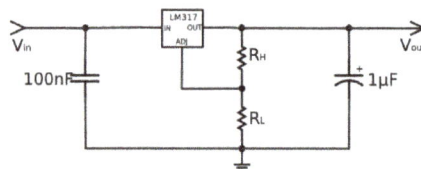

Typical electrical engineering diagram used as a troubleshooting tool

At many schools, electronic engineering is included as part of an electrical award, sometimes explicitly, such as a Bachelor of Engineering (Electrical and Electronic), but in others electrical and electronic engineering are both considered to be sufficiently broad and complex that separate degrees are offered.

Some electrical engineers choose to study for a postgraduate degree such as a Master of Engineering/Master of Science (M.Eng./M.Sc.), a Master of Engineering Management, a Doctor of Philosophy (Ph.D.) in Engineering, an Engineering Doctorate (Eng.D.), or an Engineer's degree. The master's and engineer's degrees may consist of either research, coursework or a mixture of the two. The Doctor of Philosophy and Engineering Doctorate degrees consist of a significant research component and are often viewed as the entry point to academia. In the United Kingdom and some other European countries, Master of Engineering is often considered to be an undergraduate degree of slightly longer duration than the Bachelor of Engineering rather than postgraduate.

Practicing Engineers

Belgian electrical engineers inspecting the rotor of a 40,000 kilowatt turbine of the General Electric Company in New York City

In most countries, a bachelor's degree in engineering represents the first step towards professional certification and the degree program itself is certified by a professional body. After completing a certified degree program the engineer must satisfy a range of requirements (including work experience requirements) before being certified. Once certified the engineer is designated the title of Professional Engineer (in the United States, Canada and South Africa), Chartered Engineer or Incorporated Engineer (in India, Pakistan, the United Kingdom, Ireland and Zimbabwe), Chartered Professional Engineer (in Australia and New Zealand) or European Engineer (in much of the European Union).

The IEEE corporate office is on the 17th floor of 3 Park Avenue in New York City

The advantages of certification vary depending upon location. For example, in the United States and Canada "only a licensed engineer may seal engineering work for public and private clients". This requirement is enforced by state and provincial legislation such as Quebec's Engineers Act. In other countries, no such legislation exists. Practically all certifying bodies maintain a code of ethics that they expect all members to abide by or risk expulsion. In this way these organizations play an important role in maintaining ethical standards for the profession. Even in jurisdictions where certification has little or no legal bearing on work, engineers are subject to contract law. In cases where an engineer's work fails he or she may be subject to the tort of negligence and, in extreme cases, the charge of criminal negligence. An engineer's work must also comply with numerous other rules and regulations such as building codes and legislation pertaining to environmental law.

Professional bodies of note for electrical engineers include the Institute of Electrical and Electronics Engineers (IEEE) and the Institution of Engineering and Technology (IET). The IEEE claims to produce 30% of the world's literature in electrical engineering, has over 360,000 members worldwide and holds over 3,000 conferences annually. The IET publishes 21 journals, has a worldwide membership of over 150,000, and claims to be the largest professional engineering society in Europe. Obsolescence of technical skills is a serious concern for electrical engineers. Membership and participation in technical societies, regular reviews of periodicals in the field and a habit of continued learning are therefore essential to maintaining proficiency. An MIET(Member of the Institution of Engineering and Technology) is recognised in Europe as an Electrical and computer (technology) engineer.

In Australia, Canada, and the United States electrical engineers make up around 0.25% of the labor force.

Tools and Work

From the Global Positioning System to electric power generation, electrical engineers have contributed to the development of a wide range of technologies. They design, develop, test, and supervise the deployment of electrical systems and electronic devices. For example, they may work on the design of telecommunication systems, the operation of electric power stations, the lighting and wiring of buildings, the design of household appliances, or the electrical control of industrial machinery.

Satellite communications is typical of what electrical engineers work on.

Fundamental to the discipline are the sciences of physics and mathematics as these help to obtain both a qualitative and quantitative description of how such systems will work. Today most engineering work involves the use of computers and it is commonplace to use computer-aided design programs when designing electrical systems. Nevertheless, the ability to sketch ideas is still invaluable for quickly communicating with others.

Although most electrical engineers will understand basic circuit theory (that is the interactions of elements such as resistors, capacitors, diodes, transistors, and inductors in a circuit), the theories employed by engineers generally depend upon the work they do. For example, quantum mechanics and solid state physics might be relevant to an engineer working on VLSI (the design of integrated circuits), but are largely irrelevant to engineers working with macroscopic electrical systems. Even circuit theory may not be relevant to a person designing telecommunication systems that use off-the-shelf components. Perhaps the most important technical skills for electrical engineers are reflected in university programs, which emphasize strong numerical skills, computer literacy, and the ability to understand the technical language and concepts that relate to electrical engineering.

The Shadow robot hand system

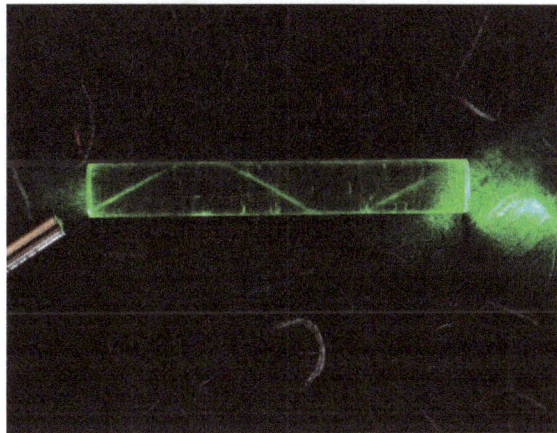

A laser bouncing down an acrylic rod, illustrating the total internal reflection of light in a multi-mode optical fiber.

A wide range of instrumentation is used by electrical engineers. For simple control circuits and alarms, a basic multimeter measuring voltage, current, and resistance may suffice. Where time-varying signals need to be studied, the oscilloscope is also an ubiquitous instrument. In RF engineering and high frequency telecommunications, spectrum analyzers and network analyzers are used. In some disciplines, safety can be a particular concern with instrumentation. For instance, medical electronics designers must take into account that much lower voltages than normal can be dangerous when electrodes are directly in contact with internal body fluids. Power transmission engineering also has great safety concerns due to the high voltages used; although voltmeters may in principle be similar to their low voltage equivalents, safety and calibration issues make them very different. Many disciplines of electrical engineering use tests specific to their discipline. Audio electronics engineers use audio test sets consisting of a signal generator and a meter, principally to measure level but also other parameters such as harmonic distortion and noise. Likewise, information technology have their own test sets, often specific to a particular data format, and the same is true of television broadcasting.

Radome at the Misawa Air Base Misawa Security Operations Center, Misawa, Japan

For many engineers, technical work accounts for only a fraction of the work they do. A lot of time may also be spent on tasks such as discussing proposals with clients, preparing budgets and determining project schedules. Many senior engineers manage a team of technicians or other engineers and for this reason project management skills are important. Most engineering projects involve some form of documentation and strong written communication skills are therefore very important.

The workplaces of engineers are just as varied as the types of work they do. Electrical engineers may be found in the pristine lab environment of a fabrication plant, the offices of a consulting firm or on site at a mine. During their working life, electrical engineers may find themselves supervising a wide range of individuals including scientists, electricians, computer programmers, and other engineers.

Electrical engineering has an intimate relationship with the physical sciences. For instance, the physicist Lord Kelvin played a major role in the engineering of the first transatlantic telegraph cable. Conversely, the engineer Oliver Heaviside produced major work on the mathematics of transmission on telegraph cables. Electrical engineers are often required on major science projects. For instance, large particle accelerators such as CERN need electrical engineers to deal with many aspects of the project: from the power distribution, to the instrumentation, to the manufacture and installation of the superconducting electromagnets.

Electronic Engineering

Printed circuit board

Electronics engineering, or electronic engineering, is an electrical engineering discipline which utilizes non-linear and active electrical components (such as semiconductor devices, especially transistors, diodes and integrated circuits) to design electronic circuits, devices, microprocessors, microcontrollers and other systems. The discipline typically also designs passive electrical components, usually based on printed circuit boards.

Electronics is a subfield within the wider electrical engineering academic subject but denotes a broad engineering field that covers subfields such as analog electronics, digital electronics, consumer electronics, embedded systems and power electronics. Electronics engineering deals with implementation of applications, principles and algorithms developed within many related fields, for example solid-state physics, radio engineering, telecommunications, control systems, signal processing, systems engineering, computer engineering, instrumentation engineering, electric power control, robotics, and many others.

The Institute of Electrical and Electronics Engineers (IEEE) is one of the most important and influential organizations for electronics engineers.

Relationship to Electrical Engineering

Electronics is a subfield within the wider electrical engineering academic subject. An academic degree with a major in electronics engineering can be acquired from some universities, while other universities use electrical engineering as the subject. The term electrical engineer is still used in the academic world to include electronic engineers. However, some people consider the term 'electrical engineer' should be reserved for those having specialized in power and heavy current or high voltage engineering, while others consider that power is just one subset of electrical engineering and (and indeed the term 'power engineering' is used in that industry) as well as 'electrical distribution engineering'. Again, in recent years there has been a growth of new separate-entry degree courses such as 'information engineering', 'systems engineering' and 'communication systems engineering', often followed by academic departments of similar name, which are typically not considered as subfields of electronics engineering but of electrical engineering.

History

Electronic engineering as a profession sprang from technological improvements in the telegraph industry in the late 19th century and the radio and the telephone industries in the early 20th century. People were attracted to radio by the technical fascination it inspired, first in receiving and then in transmitting. Many who went into broadcasting in the 1920s were only 'amateurs' in the period before World War I.

To a large extent, the modern discipline of electronic engineering was born out of telephone, radio, and television equipment development and the large amount of electronic systems development during World War II of radar, sonar, communication systems, and advanced munitions and weapon systems. In the interwar years, the subject was known as radio engineering and it was only in the late 1950s that the term electronic engineering started to emerge.

Electronics

In the field of electronic engineering, engineers design and test circuits that use the electromagnetic properties of electrical components such as resistors, capacitors, inductors, diodes and transistors to achieve a particular functionality. The tuner circuit, which allows the user of a radio to filter out all but a single station, is just one example of such a circuit.

In designing an integrated circuit, electronics engineers first construct circuit schematics that specify the electrical components and describe the interconnections between them. When completed, VLSI engineers convert the schematics into actual layouts, which map the layers of various conductor and semiconductor materials needed to construct the circuit. The conversion from schematics to layouts can be done by software but very often requires human fine-tuning to decrease space and power consumption. Once the layout is complete, it can be sent to a fabrication plant for manufacturing.

For systems of intermediate complexity engineers may use VHDL modelling for programmable logic devices and FPGAs

Integrated circuits, FPGAs and other electrical components can then be assembled on printed circuit boards to form more complicated circuits. Today, printed circuit boards are found in most electronic devices including televisions, computers and audio players.

Subfields

Electronic engineering has many subfields. This section describes some of the most popular subfields in electronic engineering; although there are engineers who focus exclusively on one subfield, there are also many who focus on a combination of subfields.

Signal processing deals with the analysis and manipulation of signals. Signals can be either analog, in which case the signal varies continuously according to the information, or digital, in which case the signal varies according to a series of discrete values representing the information.

For analog signals, signal processing may involve the amplification and filtering of audio signals for audio equipment or the modulation and demodulation of signals for telecommunications. For digital signals, signal processing may involve the compression, error checking and error detection of digital signals.

Telecommunications engineering deals with the transmission of information across a channel such as a co-axial cable, optical fiber or free space.

Transmissions across free space require information to be encoded in a carrier wave in order to shift the information to a carrier frequency suitable for transmission, this is known as modulation. Popular analog modulation techniques include amplitude modulation and frequency modulation. The choice of modulation affects the cost and performance of a system and these two factors must be balanced carefully by the engineer.

Once the transmission characteristics of a system are determined, telecommunication engineers design the transmitters and receivers needed for such systems. These two are sometimes combined to form a two-way communication device known as a transceiver. A key consideration in the design of transmitters is their power consumption as this is closely related to their signal strength. If the signal strength of a transmitter is insufficient the signal's information will be corrupted by noise.

Control engineering has a wide range of applications from the flight and propulsion systems of commercial airplanes to the cruise control present in many modern cars. It also plays an important role in industrial automation.

Control engineers often utilize feedback when designing control systems. For example, in a car with cruise control the vehicle's speed is continuously monitored and fed back to the system which adjusts the engine's power output accordingly. Where there is regular feedback, control theory can be used to determine how the system responds to such feedback.

Instrumentation engineering deals with the design of devices to measure physical quantities such as pressure, flow and temperature. These devices are known as instrumentation.

The design of such instrumentation requires a good understanding of physics that often extends beyond electromagnetic theory. For example, radar guns use the Doppler effect to measure the speed of oncoming vehicles. Similarly, thermocouples use the Peltier–Seebeck effect to measure the temperature difference between two points.

Often instrumentation is not used by itself, but instead as the sensors of larger electrical systems. For example, a thermocouple might be used to help ensure a furnace's temperature remains constant. For this reason, instrumentation engineering is often viewed as the counterpart of control engineering.

Computer engineering deals with the design of computers and computer systems. This may involve the design of new computer hardware, the design of PDAs or the use of computers to control an industrial plant. Development of embedded systems—systems made for specific tasks (e.g., mobile phones)—is also included in this field. This field includes the micro controller and its applications. Computer engineers may also work on a system's software. However, the design of complex software systems is often the domain of software engineering, which is usually considered a separate discipline.

VLSI design engineering VLSI stands for *very large scale integration*. It deals with fabrication of ICs and various electronics components.

Education and Training

Electronics engineers typically possess an academic degree with a major in electronic engineering. The length of study for such a degree is usually three or four years and the completed degree may be designated as a Bachelor of Engineering, Bachelor of Science, Bachelor of Applied Science, or Bachelor of Technology depending upon the university. Many UK universities also offer Master of Engineering (MEng) degrees at undergraduate level.

Some electronics engineers also choose to pursue a postgraduate degree such as a Master of Science (MSc), Doctor of Philosophy in Engineering (PhD), or an Engineering Doctorate (EngD). The master's degree is being introduced in some European and American Universities as a first degree and the differentiation of an engineer with graduate and postgraduate studies is often difficult. In these cases, experience is taken into account. The master's degree may consist of either research, coursework or a mixture of the two. The Doctor of Philosophy consists of a significant research component and is often viewed as the entry point to academia.

In most countries, a bachelor's degree in engineering represents the first step towards certification and the degree program itself is certified by a professional body. After completing a certified degree program the engineer must satisfy a range of requirements (including work experience requirements) before being certified. Once certified the engineer is designated the title of Professional Engineer (in the United States, Canada and South Africa), Chartered Engineer or Incorporated Engineer (in the United Kingdom, Ireland, India and Zimbabwe), Chartered Professional Engineer (in Australia and New Zealand) or European Engineer (in much of the European Union).

Some trained physicists may also choose to become Electronic Engineers.

A degree in electronics generally includes units covering physics, chemistry, mathematics, project management and specific topics in electrical engineering. Initially such topics cover most, if not all, of the subfields of electronic engineering. Students then choose to specialize in one or more subfields towards the end of the degree.

Fundamental to the discipline are the sciences of physics and mathematics as these help to obtain both a qualitative and quantitative description of how such systems will work. Today most engineering work involves the use of computers and it is commonplace to use computer-aided design and simulation software programs when designing electronic systems. Although most electronic engineers will understand basic circuit theory, the theories employed by engineers generally depend upon the work they do. For example, quantum mechanics and solid state physics might be relevant to an engineer working on VLSI but are largely irrelevant to engineers working with macroscopic electrical systems.

Apart from electromagnetics and network theory, other items in the syllabus are particular to *electronics* engineering course. *Electrical* engineering courses have other specialisms such as machines, power generation and distribution. This list does not include the extensive engineering mathematics curriculum that is a prerequisite to a degree.

Electromagnetics

Elements of vector calculus: divergence and curl; Gauss' and Stokes' theorems, Maxwell's equations: differential and integral forms. Wave equation, Poynting vector. Plane waves: propagation through various media; reflection and refraction; phase and group velocity; skin depth. Transmission lines: characteristic impedance; impedance transformation; Smith chart; impedance matching; pulse excitation. Waveguides: modes in rectangular waveguides; boundary conditions; cut-off frequencies; dispersion relations. Antennas: Dipole antennas; antenna arrays; radiation pattern; reciprocity theorem, antenna gain.

Network Analysis

Network graphs: matrices associated with graphs; incidence, fundamental cut set and fundamental circuit matrices. Solution methods: nodal and mesh analysis. Network theorems: superposition, Thevenin and Norton's maximum power transfer, Wye-Delta transformation. Steady state sinusoidal analysis using phasors. Linear constant coefficient differential equations; time domain analysis of simple RLC circuits, Solution of network equations using Laplace transform: frequency domain analysis of RLC circuits. 2-port network parameters: driving point and transfer functions. State equations for networks.

Electronic Devices and Circuits

Electronic devices: Energy bands in silicon, intrinsic and extrinsic silicon. Carrier transport in silicon: diffusion current, drift current, mobility, resistivity. Generation and recombination of carriers. p-n junction diode, Zener diode, tunnel diode, BJT, JFET, MOS capacitor, MOSFET, LED, p-i-n and avalanche photo diode, LASERs. Device technology: integrated circuit fabrication process, oxidation, diffusion, ion implantation, photolithography, n-tub, p-tub and twin-tub CMOS process.

Analog circuits: Equivalent circuits (large and small-signal) of diodes, BJTs, JFETs, and MOSFETs. Simple diode circuits, clipping, clamping, rectifier. Biasing and bias stability of transistor and FET amplifiers. Amplifiers: single-and multi-stage, differential, operational, feedback and power. Analysis of amplifiers; frequency response of amplifiers. Simple op-amp circuits. Filters. Sinusoidal oscillators; criterion for oscillation; single-transistor and op-amp configurations. Function generators and wave-shaping circuits, Power supplies.

Digital circuits: Boolean functions (NOT, AND, OR, XOR,...). Logic gates digital IC families (DTL, TTL, ECL, MOS, CMOS). Combinational circuits: arithmetic circuits, code converters, multiplexers and decoders. Sequential circuits: latches and flip-flops, counters and shift-registers. Sample and hold circuits, ADCs, DACs. Semiconductor memories. Microprocessor 8086: architecture, programming, memory and I/O interfacing.

Signals and Systems

Definitions and properties of Laplace transform, continuous-time and discrete-time Fourier series, continuous-time and discrete-time Fourier Transform, z-transform. Sampling theorems. Linear Time-Invariant (LTI) Systems: definitions and properties; causality, stability, impulse response, convolution, poles and zeros frequency response, group delay, phase delay. Signal transmission through LTI systems. Random signals and noise: probability, random variables, probability density function, autocorrelation, power spectral density, function analogy between vectors & functions.

Control Systems

Basic control system components; block diagrammatic description, reduction of block diagrams — Mason's rule. Open loop and closed loop (negative unity feedback) systems and stability analysis of these systems. Signal flow graphs and their use in determining transfer functions of systems; transient and steady state analysis of LTI control systems and frequency response. Analysis of steady-state disturbance rejection and noise sensitivity.

Tools and techniques for LTI control system analysis and design: root loci, Routh-Hurwitz stability criterion, Bode and Nyquist plots. Control system compensators: elements of lead and lag compensation, elements of Proportional-Integral-Derivative controller (PID). Discretization of continuous time systems using Zero-order hold (ZOH) and ADCs for digital controller implementation. Limitations of digital controllers: aliasing. State variable representation and solution of state equation of LTI control systems. Linearization of Nonlinear dynamical systems with state-space realizations in both frequency and time domains. Fundamental concepts of controllability and observability for MIMO LTI systems. State space realizations: observable and controllable canonical form. Ackermann's formula for state-feedback pole placement. Design of full order and reduced order estimators.

Communications

Analog communication systems: amplitude and angle modulation and demodulation systems, spectral analysis of these operations, superheterodyne noise conditions.

Digital communication systems: pulse code modulation (PCM), Differential Pulse Code Modulation (DPCM), Delta modulation (DM), digital modulation schemes-amplitude, phase and frequency shift keying schemes (ASK, PSK, FSK), matched filter receivers, bandwidth consideration and probability of error calculations for these schemes, GSM, TDMA.

Professional Bodies

Professional bodies of note for electrical engineers include the Institute of Electrical and Electronics Engineers (IEEE) and the Institution of Electrical Engineers (IEE) (now renamed the Institution of Engineering and Technology or IET). Members of the Institution of Engineering and Technology (MIET) are recognised professionally in Europe, as Electrical and computer (technology) engineers. The IEEE claims to produce 30 percent of the world's literature in electrical/electronic engineering, has over 430,000 members, and holds more than 450 IEEE sponsored or cosponsored conferences worldwide each year. SMIEEE is a recognised professional designation in the United States.

Project Engineering

For most engineers not involved at the cutting edge of system design and development, technical work accounts for only a fraction of the work they do. A lot of time is also spent on tasks such as discussing proposals with clients, preparing budgets and determining project schedules. Many senior engineers manage a team of technicians or other engineers and for this reason project management skills are important. Most engineering projects involve some form of documentation and strong written communication skills are therefore very important.

The workplaces of electronics engineers are just as varied as the types of work they do. Electronics engineers may be found in the pristine laboratory environment of a fabrication plant, the offices of a consulting firm or in a research laboratory. During their working life, electronics engineers may find themselves supervising a wide range of individuals including scientists, electricians, computer programmers and other engineers.

Obsolescence of technical skills is a serious concern for electronics engineers. Membership and participation in technical societies, regular reviews of periodicals in the field and a habit of continued learning are therefore essential to maintaining proficiency. And these are mostly used in the field of consumer electronics products.

References

- Ronalds, B.F. (2016). Sir Francis Ronalds: Father of the Electric Telegraph. London: Imperial College Press. ISBN 978-1-78326-917-4.

- Engineering: Issues, Challenges and Opportunities for Development. UNESCO. 2010. pp. 127–8. ISBN 978-92-3-104156-3.

- Manual on the Use of Thermocouples in Temperature Measurement. ASTM International. 1 January 1993. p. 154. ISBN 978-0-8031-1466-1.

- Occupational Outlook Handbook, 2008–2009. U S Department of Labor, Jist Works. 1 March 2008. p. 148. ISBN 978-1-59357-513-7.

- Charles A. Harper High Performance Printed Circuit Boards, pp. xiii–xiv, McGraw-Hill Professional, 2000 ISBN 978-0-07-026713-8

- Jimmie J. Cathey Schaum's Outline of Theory and Problems of Electronic Devices and Circuits, McGraw Hill, 2002 ISBN 978-0-07-136270-2

- Anant Agarwal/Jeffrey H. Lang Foundations of Analog and Digital Electronic Circuits, Morgan Kaufmann, 2005 ISBN 978-1-55860-735-4

- Hwei Piao Hsu Schaum's Outline of Theory and Problems of Signals and Systems, p. 1, McGraw–Hill Professional, 1995 ISBN 978-0-07-030641-7

- Gerald Luecke, Analog and Digital Circuits for Electronic Control System Applications, Newnes, 2005. ISBN 978-0-7506-7810-0.

- Homer L. Davidson, Troubleshooting and Repairing Consumer Electronics, p. 1, McGraw–Hill Professional, 2004. ISBN 978-0-07-142181-2.

- "Electrical and Electronic Engineer". Occupational Outlook Handbook, 2012-13 Edition. Bureau of Labor Statistics, U.S. Department of Labor. Retrieved November 15, 2014.

Permissions

Index

www.ingramcontent.com/pod-product-compliance
Lightning Source LLC
Chambersburg PA
CBHW061318190326
41458CB00011B/3836